PARALLEL EXPEDITIONS

PARALLEL EXPEDITIONS

Charles Darwin and the Art of John Steinbeck

Brian E. Railsback

University of Idaho Press
Moscow, Idaho
1995

Copyright © 1995 University of Idaho Press
Published by the University of Idaho Press,
Moscow, Idaho 83844-1107
Printed in the United States of America

Design by Peggy Pace

98 97 96 95   5 4 3 2 1

Library of Congress Cataloging-in-Publication Data

Railsback, Brian E.
    Parallel expeditions : Charles Darwin and the art of John Steinbeck / Brian E.
Railsback.
        p.   cm.
    Includes bibliographical references and index.
    ISBN 0-89301-177-0
    1. Steinbeck, John, 1902–1968—Knowledge—Science. 2. Literature and sci-
ence—United States—History—20th century. 3. Darwin, Charles, 1809–1882—
Influence. 4. American fiction—English literature. 7. Biology in literature.
I. Title.
PS3537.T3234Z844   1995
813'.52—dc20   94-45896
                                                                                  CIP

Permissions

For Robert DeMott,
Teacher, Scholar, Poet,
and a man whose kindness lights the whole thing up.

Contents

Acknowledgments

No one undertakes an expedition alone.

The same goes for a book. Without the help of a vast number of people, I would not have begun nor completed my *Parallel Expeditions*.

I owe the most to Sandra Lea Gary Railsback, whose love and patience prevailed over all crises. Travis Evan, Justin Gary, and Cadence Alexandra—Railsbacks all—lovingly bore my absences and kept me sane. Charles Keith and Patricia Anne Boone Railsback did what parents do: bankrolled this trip with love and money.

I had the help of two great institutions: Western Carolina University and Ohio University.

At Western, a Faculty Research Grant enabled me to travel to Steinbeck archives at Harvard, San Jose State, Stanford, and the University of Texas at Austin; I am grateful to the Office of Research and Graduate Studies—Tony Hickey, Dean, and Steve Yurkovich, Associate Dean, in particular—for this support. The Office of the Vice Chancellor for Academic Affairs, with Jack Wakeley in charge, bought me time with a Faculty Release Grant. Karl Nicholas kindly endorsed both projects. Several of my friends in the English Department provided encouragement, advice, and all those things good people will do when the ship hits a rock: Steve Eberly, Denise "My Book's A Bestseller" Heinze, Gayle Miller, Jim "Where's The Book" Nicholl, Nancy Norgaard, and Mimi "Chaos" Norton. One of my Professional Writing seniors, Jordana Stephens, proved to be an excellent assistant.

At Ohio, I owe a monstrous debt to Peter Heidtmann, who orchestrated the John Cady Fellowship that allowed me to complete the first draft of this book as a dissertation in 1989. In 1988 I had the good fortune to take Lester Marks's American Literature Seminar and Earl Knies's Seminar, "Darwin Among The Poets," at the same time; hence, I was able to make my own small inductive leap. Roy Flannagan, editor of the *Milton Quarterly*, allowed me to use his incredible computer facilities for my first draft. Edward Allen, David "Poetry Guy" Farrah, and Julie Miller gave me all the moral support I could need. And of course there was my dissertation director, Robert DeMott, to whom this book is rightfully dedicated.

Professionals in the publishing industry saw me through to the finish. Charles Fulton Campbell, Associate Editor for The National Center for State Courts, volunteered many, many long hours as a splendid copy editor. Jo Glorie, Acquisitions Editor for Paragon House, discovered this book. Susan F. Beegel, editor of the *Hemingway Review*, directed me to the University of Idaho Press, where the book was the expert hands of the staff and Toni Smith, Editor.

The disorienting world of permissions was clarified and made easy for me by Florence Eichin, of Penguin USA; Eugene Winick and Julie Fallowfield of McIntosh and Otis; and Dianne Nilsen of the Center for Creative Photography (University of Arizona).

The community of Steinbeck scholars was a beacon of wisdom and generosity, and I can name only a few here: Richard Astro, Jackson J. Benson, Donald Coers, John Ditsky, Warren French, Robert E. Morsberger, and Lewis Owens. Susan Shillinglaw, Director of the Steinbeck Research Center, opened up the Center for me and, with Ian and Nora, lent me her house for a few days. And as is so often the case in Steinbeck studies, Tetsumaro Hayashi paved the way for many research paths. I am grateful for the work and hospitality of my colleagues in Japan, who were most gracious when I visited their country to lecture in 1993; I especially wish to thank Kiyoshi Nakayama and Takahiko Sugiyama.

On March 2, 1993, Elaine Steinbeck granted me an interview at her apartment in New York City; she gave me the most memorable moment in my career by placing her husband's Nobel Prize in the palm of my hand. It reminded me that without John Steinbeck, of course, Steinbeck scholars would not exist, and neither would *Parallel Expeditions*.

Grasping for "What Actually 'Is'"

"Any critic knows it is no longer legal to praise John Steinbeck."
—Newsweek *review of* The Winter of Our Discontent, *1961*

On the night of March 17, 1989, Leslie Fiedler gave the keynote address at an international conference, *"The Grapes of Wrath,* 1939-1989: An Interdisciplinary Forum," held at San Jose State University in California. To an audience including prominent Steinbeck scholars seated near the heart of Steinbeck country, Fiedler blasted the literary reputation of the novelist in the vindictive style used by a number of critics and reviewers since the 1930s. After reading his piece, in which he identified Steinbeck as a middlebrow author for middlebrow critics, Fiedler paced about the lectern under the spotlight, shooting back answers to hostile questions from an irritated audience. Smiling slightly, eyes squinting into the light, he savored the result of his assault. Some of the "middlebrow" Steinbeck scholars left the auditorium early. Fiedler's damnation of Steinbeck was provoking, as he certainly intended it to be that night, but aside from his odd complaint that the novelist's characters are not mythic enough (comparing less favorably to such heroic figures as Scarlett O'Hara, Sherlock Holmes, and Superman), the critic had little new to say.

In a 1990 issue of *San Jose Studies* devoted to Steinbeck, the Fiedler speech is printed—a late chapter in the critical bashing of the novelist's reputation. Fiedler begins his attack in a rather wistful tone: "Why, I feel impelled to ask, has Steinbeck's reputation thus declined—so swiftly, indeed" (55). The idea that in the last twenty-five years Steinbeck has been all but forgotten by serious critics echoes Harold Bloom's 1987 contention that "his critical reputation has suffered a considerable decline" (1). Fiedler suggests that Steinbeck's demise is opposed primarily by "an ever-diminishing number of hardcore fans—chiefly Californians" (55), ignoring the numerous important scholars devoted to Steinbeck, including Jackson J. Benson, Robert DeMott, John Ditsky, Tetsumaro Hayashi, and Warren French, none of whom, with the exception of Benson, are Californians.

Fiedler marshals all of the old arguments: Steinbeck the thirties-bound provincial, the hopeless sentimentalist, the befuddled philosopher, the ambiguous politician. The critic draws from a rich tradition of literary controversy that puts Steinbeck in the same precarious canonical position as Edgar Allan Poe: writers who are, to a substantial number of scholars, unaccountably popular.

Fiedler's accusation that Steinbeck is "maudlin" and "sentimental" has quite an ancestry among the novelist's literary executioners. Since this is the most frequent charge leveled against Steinbeck's works, we might study it as a line of contradictory denigration typical of the novelist's detractors. In 1937, *New York Times* critic Ralph Thompson described *Of Mice and Men* as "a grand little book, for all its ultimate melodrama" with a plot that "probably sounds like sentimental truck, and in a way it is" (L15). Later, he finds *The Red Pony* to be a "sentimental tragedy on the order of 'Of Mice and Men'" (L21). Orville Prescott picks up the baton in 1945, observing the sentimentality of *Cannery Row.* Even a novel as dark and bitter as Steinbeck's last, *The Winter of Our Discontent,* makes an anonymous reviewer in the *Times Literary Supplement* remark that it is "sentimental and rather trivial at heart" (413).

In contradiction to these accusations of sentimentality, many Steinbeck denigrators fault his portrayal of earthy characters—a trait of tough realism that goes back to Rebecca Harding Davis and Stephen Crane. The denizens of *Cannery Row,* those "run-of-the-mill people," surely irritate Prescott: "The general atmosphere is one of biological benevolence, a sort of beaming approbation for human activities conducted on an unthinking level far below the demarcation line of pride, honesty, self-respect and accomplishment" (L17). That same novel suffers a similar attack from the *Times Literary Supplement;* the reviewer complains about Steinbeck's "feeling for the grotesquely human or the humanly grotesque" (521). In his damnation of Steinbeck's Nobel Prize, Arthur Mizener rebukes the author's adoration of primitives (and disdains the sentimentality as well). Fiedler is appalled by the "reprehensible" characters of *The Grapes of Wrath:* "Examples include not only grossness, blasphemy, and a contempt for literacy, but habitual drunkenness, loveless tom-catting, petty thievery, and finally mindless violence, from bar-room brawling to wife-beating and murder" (58). We might suspect Prescott, Mizener, or Fiedler of wishing Steinbeck would be *more* sentimental and less realistic, would, for example, clean up the Okies and make them more palatable for literary tastes.

In his Steinbeck edition for the Modern Critical Views series, Bloom includes a 1947 article by Donald Weeks which castigates the novelist both for his sentimentality *and* for the crude practices of his characters. Weeks is one of the rare critics before Peter Lisca, Lester Marks, or Richard Astro to understand something of Steinbeck's scientific, nonteleological philosophy, yet he writes it off as the novelist's attempt "to rationalize his sentimentality" (8). He ignores Steinbeck's plan of seeking truth, or "what actually 'is,'" by an inductive, scientific methodology. Instead, Weeks reveals his disgust for Steinbeck's low characters: "I want to call the celebration of impulse in Steinbeck the philosophy of the wino" (13). Weeks sums up the contradictory line of the critics of Steinbeck's sentimentalism. "When Steinbeck gets sentimental, life becomes warm, beautiful, satisfying," he writes (13). Then he attacks the author's work because the "later novels have been so inhuman, so deeply contemptuous, I would say, of men and women" (15). Is Steinbeck, then, a paradox—a tough sentimentalist? Weeks concludes with three sentences that underscore his confusion: "I thought I knew what was the matter with Steinbeck. I could be wrong. I want to be wrong" (17).

Among Steinbeck's harshest critics, Weeks does not wander about slightly baffled on his own. In his introduction to the Modern Critical Views edition, Bloom begins by noting that *The Grapes of Wrath* "is a very problematical work, and very difficult to judge" (1). He has very little else to say about it, noting its popularity with "liberal middle-brows" (1), and saying more about Hemingway, Whitman, and Emerson than Steinbeck. Uneasy about his rereading of *The Grapes of Wrath,* Bloom concludes ambiguously, "One might desire *The Grapes of Wrath* to be composed differently, whether as plot or as characterization, but wisdom compels one to be grateful for the novel's continued existence" (5). Also somewhat confused, Fiedler calls *The Grapes of Wrath* a time-bound socialist book, a work for "the Cultural Commissars in Moscow" (56). He asserts that *Grapes* is a failure by modernist standards (which Fiedler himself has abandoned) and is a novel with characters outrageous to contemporary sensibilities, somewhere between high art and pop culture, a place Fiedler does not find "easy to define" (61). He concludes that the book is a failed pastoral (Marxist pastoral?), perhaps a Western-style Oedipal play—ultimately a "problematical, middlebrow book" (64).

We could continue for many pages observing the problems Steinbeck has caused his reviewers and critics, exploring contradictory lines of attack such as the wish that the novelist would not be so sentimental and, at the same time, not be so cold. Indeed, Steinbeck's trouble with the critics has been so great that it has become a tradition in the scholarly conversation about him to begin by lamenting or agreeing with the harsh reception of his works.

In the first substantial book about Steinbeck since Harry Thornton Moore's pioneering 1939 study, *The Novels of John Steinbeck*, Peter Lisca began the tradition of examining Steinbeck's critical treatment with the publication of *The Wide World of John Steinbeck* in 1958. His introduction, "The Failure of Criticism," is an excellent review of Steinbeck's detractors from the forties and fifties. Lisca's thesis is that Steinbeck's works "have been both accepted and rejected on sociological rather than aesthetic terms" (3). He also is the first to make the crucial observation that Steinbeck's biological views have been largely misunderstood, beginning with Edmund Wilson's extremely influential essay, "The Californians: Storm and Steinbeck" (revised as "John Steinbeck" in Wilson's *The Boys in the Back Room* [1941]). Many critics following Wilson have reacted negatively to what they feel is Steinbeck's animalism and his befuddled biological perspective, what Fiedler calls "eccentric biologism" (55). Lisca concludes that Steinbeck's reputation suffers because "he seldom *has* been taken seriously insofar as this seriousness demands formal analysis of [his] works" (19).

Three years after Lisca, Warren French wrote in the preface of his *John Steinbeck* that the novelist "is not critically fashionable today" (7). French notes that the "critical ill-logic" has been to rate Steinbeck by judging him against the decline of his works since *The Grapes of Wrath,* rather than recognizing his other achievements of the thirties. Steinbeck's tendency to use allegory and his transcendentalism, French asserts, have given him additional trouble with twentieth-century critics uncomfortable with these unmodern modes. Like Lisca, French believes that Steinbeck's nonteleological viewpoint has contributed to the difficulty of assessing his work. Perhaps the most important statement in this preface concerning the critics is French's belief that "some do not like Steinbeck because he does not see the world as they see it and does not tell them what they wish to hear" (7). Regarding Steinbeck's view of human *as* animal, French's observation is particularly important.

Richard Astro's excellent 1973 study of the influence upon Steinbeck by his marine biologist friend, Ed Ricketts, *John Steinbeck and Edward F. Ricketts: The Shaping of a Novelist,* begins by discussing the critical misunderstanding of Steinbeck's biological views. Astro believes that critics have found Steinbeck's philosophy confused or inadequate because they lack the biographical facts or the scientific background necessary to understand it. He writes that "most critics carry out their investigations within the constricted framework of literary patterns and traditions. . . . [T]his approach may be sufficient, but with Steinbeck it is disastrous" (4). Astro's point is perfectly illustrated by R. W. B. Lewis' 1958 essay, "The Picaresque Saint," in which he notes "a certain failure of artistic promise" in Steinbeck's work (144). Steinbeck, he continues, has pursued social questions in *In Dubious Battle* "in artistically unfruitful directions" because he has "talked about nodes and nuclei and organisms and cells, wasting his poetic vein on scientific and social-scientific abstractions" (145). Lewis considers even *The Grapes of Wrath* a failure because the "quasi-scientific interests" get in the way of a story about human beings (147). However, Lewis never bothers to define what Steinbeck's scientific interests are; he merely objects to them, perhaps because he does not want either to accept or to understand the novelist's view.

A very lucid account of the critics' failure to understand Steinbeck, and in particular his biological perspective, is contained in Jackson J. Benson's 1988 book, *Looking for Steinbeck's Ghost.* Benson has written the most important contribution to Steinbeck scholarship, 1984's monumental authorized biography of the author, *The True Adventures of John Steinbeck, Writer,* so his recognition of the importance of biology in Steinbeck's work and the devastation caused by critics who fail to recognize it is not surprising. Although Benson believes that Steinbeck wrote "a few very bad books," he is upset by his discovery that "much of the criticism of Steinbeck's work was just gratuitously nasty" (183, 184). Among the reasons Benson cites for this nastiness are that Steinbeck is a popular writer, has a sense of humor, and is willing to experiment (he "refused to write the same book [*The Grapes of*

Wrath] over and over again" [184]). But the greatest problem Steinbeck has with the critics is "his politics, or lack of them" (184). The novelist has been squeezed from the right by critics who call him a Marxist and from the left by "the snobbish disdain" of liberal intellectuals (184). Once again, the critical view of Steinbeck's books—this time from a political perspective—is contradictory. "Wrapped up in their own political emotions," Benson writes, "reviewers on the right or the left seldom paid much attention to what Steinbeck was actually saying" (185). An important element of this inability to read Steinbeck's work beyond a political framework is a blindness to his ecological perspective: "When he talked about the human 'species' and the need to live in harmony with the whole of nature and the need to adapt . . . he might as well have been talking Martian, as far as most literary critics were concerned" (196-97).

What links Steinbeck's detractors, then, is that they all fail to understand—or even recognize—the unifying subject of nearly all of Steinbeck's work: his biological perspective. This lack seems particularly painful in Bloom and Fiedler, who, writing in the eighties, have had access to important observations of Steinbeck and biology made by Lisca, Astro, and Benson. In fact, Bloom and Fiedler need look no further than Edmund Wilson, whose work they surely know. Wilson's essay on Steinbeck identifies the constant in the novelist's work: "his preoccupation with biology" (42). Although admonishing Steinbeck, finding journalism, theatricalism, and tricks in his work, Wilson recognizes that the author's scientific, "unpanicky scrutiny of life" is characteristic of a "first-rate" mind (53). He notices that biology unifies the diverse range of works from *Tortilla Flat* to *The Grapes of Wrath*, from the "comic idyl" to the "propaganda novel" (41, 42). He knows Steinbeck "is a biologist in the literal sense," concerned with "the processes of life itself . . . the predatory appetite and the competitive instinct that are necessary for the very survival of eating and breeding creatures" (42, 44). However, Wilson goes on to observe that Steinbeck does not write of the "thoughtful, imaginative, constructive" aspects of humanity (44). "This animalizing tendency," Wilson writes, "is, I believe, at the bottom of his relative unsuccess at representing human beings" (48). Steinbeck's philosophy simply is "not enough" for Wilson (51), who cannot accept the world view he has seen in Steinbeck's work.

Steinbeck's "animalizing tendency" really becomes a greater problem for his detractors than for the author. With the exception of

Wilson, Steinbeck's attackers have little understanding of the impact of science on his works. These critics, thoroughly schooled in aesthetics, modernism, and humanism, are at sea when encountering John Steinbeck's biological perspective; the points of the compass have been changed and the critical navigators, looking for more familiar landmarks, become disoriented. Consequently, Steinbeck's position in the canon slips as critics such as Bloom and Fiedler compare him to Hemingway and Faulkner, other thirties writers and Nobel Prize winners. And an anonymous reviewer for *Time*—blasting the Nobel Prize award—writes that Steinbeck uses the Salinas Valley as his "strip-cartoon Yoknapatawpha County" ("Wrapped and Shellacked" 41).

These reviewers and scholars do Steinbeck's literary reputation a terrible disservice, firing crushing broadsides without a fair understanding of the target. In Steinbeck's "low" characters, his detractors conduct a futile search for human nobility, for tales of the human heart warring against itself, or for stoicism in the face of an indifferent or hostile universe. But, as a thorough study of Steinbeck's approach will show, a strictly human-centered story rooted only in humanist philosophy should not be looked for in Steinbeck until his departure with *East of Eden.*

Steinbeck's contribution to American literature is unique, for he offers dramatizations of biological principles—the "what actually 'is'" (*log* 139) in our interrelations with the natural world—that cut through Christianity, humanism, economics, sociology, and politics. Unlike the famous practitioners of literary naturalism a generation before him, such as Theodore Dreiser, Jack London, or Frank Norris, Steinbeck entirely bypassed the politically and philosophically charged biological viewpoint of Herbert Spencer (there is no evidence that the novelist ever read Spencer). Steinbeck wrote not in the tradition of the liberal arts, but more in the tradition of the sciences. He tried to break away from artistic subjectivity and to perceive and present the human species from the objective, sometimes callous view of a laboratory technician. Small wonder that humanist critics, from Orville Prescott to Leslie Fiedler, find something awry with an author who knocks the species from its self-appointed position in center stage.

Steinbeck himself recognized the trouble he gave his critics and occasionally the mean-spirited assessments directed at him. In January of 1963 he wrote to his friend and editor, Pascal Covici, about Mizener's assault on the novelist's Nobel Prize: "Whatever his virtues are, and he must have some, he is certainly bad mannered" (Fensch, *Steinbeck and Covici* 230). He also commented on the familiar charge Mizener made that Steinbeck's works are sentimental: "Now, does he say that there are no relationships between people which he would call sentimental? If this is his meaning, he is wrong or inexperienced or unobserving" (229). Steinbeck objected to the modern trend, represented by Mizener, "that the writer may mention only the cowards and the insensitives but ignore the heroes" and added, "What it boils down to is that everything exists" (229-30). In other words, Steinbeck is after the *whole* range of human experience, including the evil, the humorous, and even the sentimental. Indeed, Steinbeck's biological perspective compels him to pursue "what actually 'is,'" what his friend Ed Ricketts called the "toto" picture.

The author was also aware that his scientific view of human beings confounded reviewers. In another response to his critics, Steinbeck wrote, "It is not observed that I find it valid to understand man as an animal before I am prepared to know him as a man. It is charged that I have somehow outraged members of my species by considering them part of a species at all" ("A Postscript from Steinbeck" 307). Nothing could be more unfortunate for Steinbeck than the fact that so many of his judges miss the most important signature of his work: the view of humans as a species.

Such critics miss the most crucial level of meaning in Steinbeck's art. He wrote that his work could be understood on a number of levels. Writing to Covici, he asserts of the *The Grapes of Wrath* that "there are five layers in this book" (Fensch, *Steinbeck and Covici* 20). Of his next book, *Sea of Cortez: A Leisurely Journal of Travel and Research* (written with Edward F. Ricketts), he writes that "there are four levels of statement in it and I think very few will follow it down to the fourth. . . . [I]t is a new kind of writing. . . . I [have] found a great poetry in scientific writing" (Fensch 31). His works can indeed be understood on many levels, from realism to allegory. However, the most profound theme running throughout almost everything he wrote, perhaps that fourth level that he put into *Sea of Cortez*, is the scientific, biological one. At the deepest level is Steinbeck's desire to understand the human *as* a species, *as* an animal; this is his original vision, and the one that beneath all of the apparently contradictory technique, politics, and philosophy runs like a steady subterranean river. Here one finds the cohesive, intelligent world view that so many of his critics have missed or refused to understand.

❊ ❊ ❊

That an American novelist often thought of as a proletarian writer should have much in common with a genteel Victorian naturalist might seem incredible. However, Steinbeck was not only directly and indirectly influenced by Charles Darwin's work, but the author's quest for a true look at humanity leads him on an expedition in art that parallels in method and result the naturalist's expedition in biology. Both Steinbeck and Darwin desire to remain objective, work inductively, and persevere courageously. The attempt to break out of "humanity" and see *Homo sapiens* is a difficult and dangerous ideal: To adopt Steinbeck's and Darwin's view is to wipe away the towering preconceptions that place the human apart from other animals, and both men take a beating from their critics. Darwin takes fire from the creationists, and Steinbeck gets it from the humanists. Ironically, many of Steinbeck's detractors would probably join the modern chorus of praises for the naturalist who largely originated the view Steinbeck dramatizes.

To be sure, a number of critics have sensed something Darwinian about Steinbeck's work. Lisca, for example, notes Steinbeck's interest in survival, and writes that "this Darwinian element" exists even in the first novel, *Cup of Gold* (*Wide World* 182). Joseph Fontenrose notices some hints of "social Darwinism" in *The Log* (95). Lester Marks, in *Thematic Design in the Novels of John Steinbeck* (21, 25), recognizes Steinbeck's allusions to Darwin in *The Log from the Sea of Cortez,* and Astro mentions in passing that Steinbeck and Ricketts operated as "quasi-Darwinian naturalists" during their expedition to the Sea of Cortez (*Steinbeck and Ricketts* 30). Writing about *The Grapes of Wrath,* Louis Owens observes that Steinbeck's "interest is in man as a product of Darwinian evolution" (13), while R. S. Hughes notices "Darwinian themes" in an obscure Steinbeck story, "The Time the Wolves Ate the Vice-Principal" (76). And Bobbi Gonzales and Mimi Gladstein comment that the character of Camille in *The Wayward Bus* is "a case study of Darwinian naturalism" (162). Two interesting comments concerning Darwin and Steinbeck occur in essays for *John Steinbeck: Asian*

Perspectives. Jin Young Choi, discussing Steinbeck studies in Korea, notes that a dissertation by Pong Shik Kang finds the novelist's characterization is weakened by the "biological viewpoint" which suggests that "the sad pictures of man's inhumanity to man are seen as no more than a phase of the biological struggle for survival in the Darwinian order" (22). Hiromasa Takamura sees the play version of *Of Mice and Men* as "a microcosm of the Darwinian world where the strong destroy the weak and where the weak are forced to play undesirable roles to survive" (94). Yet all of these observations of the Darwinian in Steinbeck are relatively insignificant remarks in the books and essays noted above; Darwin's name appears once or twice, perhaps in a paragraph or two, and the subject is dropped. Darwinian elements in Steinbeck's art, however, deserve much more than a glance.

A Darwinian view of Steinbeck opens up a number of important aspects of his fiction and nonfiction prose, from *Cup of Gold* to *America and Americans,* revealing a unifying element beneath what appear to be oscillations between political sides, between hope and despair, and between sentimentalism and realism. Such a view shows the determined effort by Steinbeck to expose preconceptions as the bases of destructive dreams and self-delusions, revealing prejudice to be his bane as it is Darwin's. It shows that Darwin and Steinbeck both prize truth and decide that the best way to find it is by the inductive method of science. It shows that the result of that method, for them, is a realization of the human place as one not above but among the other species of the world, with no place for anthropocentrism or selfishness. It shows, above all, that an author who does not merely adopt social Darwinism but instead dramatizes the actual biological principles must work differently, must present characters in a particular way, and must experiment with narrative forms that underscore biological realities.

This book will examine these assertions by following the biological substructure of Steinbeck's work, particularly as it relates to Charles Darwin, clarifying the source of Steinbeck's biological philosophy of human development and progress. The three most influential critics of the author's scientific ideas—Lisca, Astro, and Benson—have done a splendid job examining the crucial influence of Ed Ricketts upon Steinbeck's views. However, the tendency has been to suggest Ricketts as *the* source for Steinbeck's interpretations of biology. Although Astro concedes that to say the novelist's interest in science and biology came directly from Ricketts "is to distort the facts," he nevertheless writes that "the train that killed Ricketts set off a series of reactions that helped kill Steinbeck as a serious novelist" (*Steinbeck and Ricketts* 228). Looking at Darwin's influence illuminates Steinbeck's own intellectual path. While Ricketts's ideas concerning nonteleology influence the novelist's methodology for finding truth, Steinbeck parallels Darwin more closely than Ricketts because Darwin's theories of competition and evolution provided Steinbeck with a biological basis for progress not found in Ed's nonteleological world view.

Chapter 2 of this study examines Steinbeck's access to Darwin, directly and indirectly. Of particular interest are the books Steinbeck read up to the time of his and Ricketts's 1940 collecting expedition in the Sea of Cortez, especially the discussions of Darwin's ideas in such works as Jan Christian Smuts's *Holism and Evolution* and John Elof Boodin's *Cosmic Evolution*. This chapter focuses also on Steinbeck's reading of Darwin's journal of the *Beagle* expedition and the novelist's view of the trip to Cortez as one similar in methods and objectives to Darwin's. Because Darwin originated many of the ideas Steinbeck and Ricketts pursue, the novelist writes of the naturalist with enthusiasm and admiration.

The rest of this study will explore how Darwinian concepts are applied in Steinbeck's works. Chapter 3 examines the author's placement of *Homo sapiens* in the whole—a viewpoint in defiance of theological and philosophical preconceptions that separate humans from nature. This holistic conception of the world, made popular by Darwin, permeates Steinbeck's style and motifs. From a narrative stance that equates humans with nature through personification, anthropomorphism, theriomorphosis, and repeated images such as a preoccupation with primitive settings, Steinbeck seeks to show that civilization and ideology often veneer what Darwin called our "lowly origin."

Chapter 4 considers Darwin's and Steinbeck's cold and even brutal view of sex. The novelist has been accused of being a misogynist, especially because of such works as "The Murder" and *The Wayward Bus*, but biographical evidence does not support a picture of Steinbeck as a misogynist writer, even though he suffered a bitter divorce from his second wife in 1948. However, Darwin's theory of sexual selection, in which the male is seen as superior to the female due to the male's fierce mating competition, is very disturbing when applied to humans. The extension of sexual selection to *Homo sapiens* creates problems for both Darwin and Steinbeck. Yet, coupled with Steinbeck's disgust for the illusions of a commercialized society, the biological view leads him to a theme of cultural oppression in *The Wayward Bus* that forecasts the feminist argument of Naomi Wolf's 1991 book, *The Beauty Myth*.

Chapter 5 details Steinbeck's and Darwin's concern with defining humanity. The novelist and the naturalist come upon strikingly similar definitions of what it is to be human. For both, the highest intellectual powers adhere to unprejudiced, nonanthropocentric observation and inductive reasoning. Such enlightened thought requires bravery, humility, and, above all, sympathy. Steinbeck's books, from *Cup of Gold* to *The Grapes of Wrath*, represent the author's own kind of evolution, as he searches for and ultimately finds a protagonist with those qualities which extract the human from *Homo sapiens*.

A close rereading of *The Grapes of Wrath* from a Darwinian perspective concludes my analysis. This great work encompasses nearly every parallel between Darwin and Steinbeck covered in this study, although a trek through Steinbeck's prose will reveal Darwinian tendencies in nearly every piece the author wrote. The two men move in surprisingly similar ways to arrive at a similar conclusion, the humbling fact that we are, despite all wishful thinking, animals. That Steinbeck undertook an expedition that paralleled Darwin's must be understood if we wish to fully appreciate the novelist's unique perspective, to move beyond the legacy of critical misinterpretation summarized by Leslie Fiedler's 1989 speech, and to discover what seeing human as *Homo sapiens* means to John Steinbeck, to Charles Darwin, and perhaps to us as well.

2

Homage to the "Older Method"

"The Darwinian period of thought, in its infancy in 1835 even in Darwin's mind, is now at full tide in the mind of every naturalist in the world."
—Henry Fairfield Osborn, preface to Galapagos, World's End, *1924*

The Sea of Cortez, a rift in the side of the lower North American continent, slices upward until it meets the arid remnant of the Colorado River. A lonely sea with long stretches of uninhabited shoreline, its waters are alive with marine life while the land remains dead, a silent witness to mirages and, occasionally, freakish electrical storms. As a remote area of burned country and vigorous ocean life, as a place of mystery, it is reminiscent of the Galapagos Islands. Cortez was for John Steinbeck what Galapagos was for Darwin: a pristine panorama of the natural world, perfect for the illustration of profound interpretations of biology.

Steinbeck clearly realized the parallels between his and Ed Ricketts's expedition aboard the *Western Flyer* and Darwin's expedition aboard the *Beagle*. Yet Steinbeck's recognition of Darwin's achievement goes beyond the superficial comparison of voyages. Evolutionary theory underscores the power of inductive, objective reasoning; it represents the first thoroughly holistic view of *Homo sapiens* and nature, successfully challenging theological and philosophical preconceptions. As a novelist fascinated by evolutionary progress, natural selection, and the inductive search for truth—all things of central importance to his artistic vision—Steinbeck's reverence for Darwin, so apparent in his most important treatise, *The Log from the Sea of Cortez,* seems only natural. However, Charles Darwin and evolution first came to the novelist long before the *Western Flyer* left Monterey for the gulf in 1940.

PREPARATIONS FOR A TRIP TO THE SEA OF CORTEZ

Trying to determine precisely Steinbeck's initial contact with Darwin is impossible, but he had a formal dose of evolutionary theory when he took general zoology at Stanford's Hopkins Marine Station in the summer of 1923. Of all the important naturalists and biologists with interpretations of Darwinian theory that Steinbeck learned from, the first was William Emerson Ritter. Richard Astro has shown that Steinbeck's professor at Hopkins was C.V. Taylor, who studied at Berkeley under Charles Kofoid, who in turn "undoubtedly had come under the influence of the ideas of William Emerson Ritter," so that "Ritter's ideas were transmitted via Kofoid and Taylor to the impressionable Steinbeck" (*Steinbeck and Ricketts* 44). From Astro's interview of Hopkins professor Rolf Bolin in 1970, we know that Ritter's concept of superorganicism was what Steinbeck remembered most from the course years later (44). Further, Benson suggests that Steinbeck may have read some of Ritter's *The Unity of the Organism, or The Organismal Conception of Life* (*True Adventures* 240).

Astro and Benson have thoroughly discussed the effect upon Steinbeck of Ritter's holistic conception of organismal unity. Ritter's philosophy of the natural world emphasizes the interrelation of the whole of nature and all of its parts; the whole not only depends upon the parts from which it is made but also determines the direction of the parts. This holistic view of nature became a major theme in most of Steinbeck's work, and, as Benson accurately observes, this "perception was further refined and modified in discussions later with Ricketts, who held a similar view, and by other reading in evolutionary philosophy" (*True Adventures* 241).

If Ritter's book is any indication, Steinbeck's introduction to the methods and views of Darwin were positive. In discussing great naturalists who possess a wider view than modern specialized scientists, Ritter writes, "three names that stand out with mountain-like conspicuousness among those who in modern times have made the idea of evolution a household possession [are] Lamarck, Darwin, and Wallace" (1: 75). Ritter accuses his contemporary biologists of taking their subjects so compartmentally that they resemble "pre-Darwinian naturalists" (1: 76). He applauds the inductive method of Darwin and examines the naturalist's discovery of evolution—how the intense curiosity of a youth aboard the *Beagle* became, after a long consideration of gathered information, a theory years later. Ritter praises Darwin's creativity and honesty: "The genuiness of the individual, the personal, the unique character of mental life and mental creation can hardly be more strikingly illustrated than by such cases [as] this of Darwin's when the conception, the hypothesis, is kept to one's self so long in order 'to prove' whether it is 'true' or not" (1: 227). Ritter's view is fairly close to Steinbeck's own appraisal of Darwin in *The Log*. If, as Astro and Benson suggest, Steinbeck's introduction to biological concepts and the philosophy of Ritter in Taylor's zoology class had a lasting effect on the author, then the novelist's positive view of Darwin had been set before he even met Edward F. Ricketts. Furthermore, Ritter's reflections upon Darwin are part of a trend of admiration that can be seen in other evolutionary thinkers Steinbeck read.

Besides Taylor, the other Stanford professor who influenced Steinbeck was Harold Chapman Brown, who taught a history of philosophy course that Steinbeck took in the winter of 1924. Benson writes that Steinbeck attended Brown's lectures regularly and met with the professor socially (*True Adventures* 235). Although Brown is no Darwinist, he shows an astute awareness of evolutionary theory: "Variation within species seems to be the necessary condition of biological evolution," he writes in a 1920 article entitled "The Problem of Philosophy," "but some variations are monsters and accomplish nothing" (285). His emphasis on holism and scientific truth key into the whole of Darwinian thought. Discussing the near impossibility of achieving a "synthetic conception of the universe," an understanding of how all the parts in a complex whole function, Brown writes in the same article, "The nearest approach to such an idea is expressed by evolution, but evolution, strictly understood, means nothing more than descent tracing and is but a way of confirming the same sort of unity of the whole through time" (286). Brown argues against the "narrowed vision" of closed philosophical systems, particularly those which ignore science. "There are no philosophic truths, but only scientific truths," he asserts, noting also that in science "the fundamental category is description" (285).

In a 1925 article, Brown advocates a kind of holism, a view of "cosmic integration," a "genuine materialism," which "asserts the continuity of the processes of life and mind with those of physical matter and grants that the same method of analysis is everywhere valid" ("The Material World—Snark or Boojum?" 214). If, as Benson believes, the two essays in the *Journal of Philosophy* represent some of the subject matter of Brown's lectures, then the professor's influence is very significant. Brown lays the seeds of many ideas that flourish in Steinbeck's work: the emphasis on science as a way to truth, the need for a wider vision, and the integration of the human species in a much larger whole.

The holistic philosophy of Ritter and Brown appears in Steinbeck's early fiction. The stories Steinbeck wrote during his last attempt to finish at Stanford in 1924-25 (he never took a degree there) indicate the growing importance in his work of connecting humans to nature. Stories such as "The Days of Long Marsh" (circa 1924) contain stylistic devices that connect characters to the land or to their primitive past. Longer manuscripts he was working on in the late twenties, "The Green Lady" (an early draft of *To a God Unknown*) and *Cup of Gold,* also demonstrate attempts to place humans in the natural whole.

This tendency is especially clear in "The Green Lady," which was first begun as a short story and then an unfinished play by one of Steinbeck's Stanford friends, Webster "Toby" Street (Benson, *True Adventures* 108). According to Benson, Street passed his problematic play onto Steinbeck for completion. In 1928, Steinbeck began rewriting "The Green Lady" as a novel; Street's story line of a California rancher, Andy Wane, who confuses his love for his land with his incestuous love for his daughter, Susie, gave Steinbeck "ideas that fit very nicely into his own strange mixture of the mystical and the biological" (139). Through narrative technique, Steinbeck's version of "The Green Lady" implicitly emphasizes the holistic possibilities of the protagonist's connection with the land.

As much as Steinbeck's experience at Stanford and Hopkins Marine Station may have given him some of the foundations upon which he built his mature work, his education really began at Cannery Row with Ricketts—who was less a teacher and more a fellow classmate. For all of the substantial influence Ricketts had upon Steinbeck, it is important to remember that the author came with some scientific notions of his own. In an unpublished letter, a mutual friend of Steinbeck and Ricketts, Richard Albee, writes in 1979 to critic Robert DeMott, "I have begun to bridle a bit when scholars seem to give Ed Ricketts more credit than I *know* he warrants in the development of John—either mind or writing." Albee was a student at UCLA who introduced John and Ed to ideas of professor John Elof Boodin that were crucial to the development of Steinbeck's notion of the phalanx, or group man. Albee contends that "John was very much his own man, and no creation of Ed's." Indeed, in his exhaustive reconstruction, *Steinbeck's Reading: A Catalogue of Books Owned and Borrowed,* DeMott concludes that "it is erroneous to think Steinbeck oscillated solely in Ricketts' rainbow. . . . As a scientist, Ricketts used his sources differently from the way Steinbeck did as a novelist" (xxvii). As Albee writes, "John's reading should indicate clearly enough that he drew his knowledge from many, many sources independent of Ed or me or anyone in particular." These observations about Steinbeck's independent use of material are especially important when considering the impact of Darwin upon his works.

Nevertheless, the Steinbeck/Ricketts relationship is one of those rare ones in American literature where the usually lonely, individualistic life of the artist is suffused by the mutual contemplations of another's mind. In his poignant memoir to his friend, "About Ed Ricketts," Steinbeck wrote that he met Ricketts in a dentist's waiting room in the fall of 1930. "Knowing Ed Ricketts was instant," Steinbeck writes. "After the first moment I knew him, and for the next eighteen years I knew him better than I knew anyone, and perhaps I did not know him at all" (xiii). Unlike Steinbeck, Ricketts was a professional biologist. He had studied at the University of Chicago and in 1923 opened a biological supply house in Pacific Grove, California, with a partner, A. E. Galigher (Astro, *Steinbeck and Ricketts* 5). By the time Steinbeck met Ricketts, Pacific Biological Laboratories, Inc., was in a small building by the seashore on Monterey's Cannery Row (Steinbeck, "About Ed" xiii). Ricketts, sole owner of the lab at Cannery Row, collected small mammals and a large quantity of various marine creatures that he preserved and sold to supply companies and schools for study and dissection. The lab is faithfully reproduced in many of Steinbeck's works, and *Cannery Row* in particular.

Steinbeck had arrived at nearby Pacific Grove in 1930 in abject poverty with his bride, Carol Henning. During the thirties, Ed's lab became a refuge for Steinbeck, a place to read, discuss ideas, and drink. Ricketts and Steinbeck had planned to collaborate on a study of the littoral (that zone of shoreline inside the marks of high and low tide) north of San Francisco but, as Astro notes, the work was never completed (*Steinbeck and Ricketts* 10). The next collaborative project was a success, resulting in *Sea of Cortez: A Leisurely Journal of Travel and*

Research with a Scientific Appendix Comprising Materials for a Source Book on the Marine Animals of The Panamic Faunal Province, issued in 1941 by Viking Press, Steinbeck's publisher. Steinbeck focused on the narrative journal, while Ricketts concentrated on the extensive scientific appendix. After Ricketts's death, Viking published Steinbeck's narrative, *The Log from the Sea of Cortez,* along with "About Ed Ricketts" in 1951. A third collaborative effort, similar to *Sea of Cortez,* was to concern collecting expeditions Ricketts made to Vancouver Island and the Queen Charlottes in 1945 and 1946. However, Steinbeck's interests and fortunes had taken new directions by this time, and he may have been reluctant to become involved in a new collaboration, as Astro suggests (*Steinbeck and Ricketts* 22).

This third venture was never completed, but not because Steinbeck ever refused it. On May 7, 1948, Ed Ricketts drove his Packard onto a blind railroad crossing close to the lab and was hit by the Del Monte Express. He died four days later, before Steinbeck, rushing from his home in New York, could see him. Thus ended Steinbeck's most important literary relationship. The two men's thoughts were very close, as Steinbeck writes: "Very many conclusions Ed and I worked out together through endless discussion and reading and observation and experiment. We worked together, and so closely that I do not now know in some cases who started which line of speculation since the end thought was the product of both minds. I do not know whose thought it was" ("About Ed" xliii). In a 1958 interview for Australian television, an interviewer's comment about Ricketts caused Steinbeck to become suddenly serious in an otherwise playful exchange. A deep feeling for his friend and a sense of pain at his death still haunted Steinbeck's words ten years after the accident: "He was my partner for eighteen years—he was part of my brain. At one time a very eminent zoologist said that the two of us together were the best zoologists in America, and when he was killed I was destroyed" (Fensch, *Conversations* 68). Despite the importance of this friendship to Steinbeck, his observation that the thinking on Cannery Row "was the product of both minds" suggests that the novelist had some ideas of his own.

Many of the scientific books Steinbeck encountered between 1930 and 1936, however, came from Ricketts's tiny library (they were all destroyed in a fire in November of 1936), and, clearly, these works profoundly affected Steinbeck's art. Virginia Scardigli, an anthropology student who often visited the lab, describes the library as some shelves on the wall in Ed's room with "not more than 50 volumes" of scientific books. Steinbeck was ever ready to read them for, at the time he met Ricketts, the novelist began "the most intensive reading program he had ever undertaken" (DeMott, *Steinbeck's Reading* xxvi). Martin Bidwell, a struggling novelist who visited the Steinbecks at their Pacific Grove home in 1933, noted the piles of magazines and books there (Benson, *True Adventures* 282-83). What could not be found at home or in Ed's lab, was borrowed at the public library in Pacific Grove. By examining some of these books, we can see where Darwin's name might have come up again and again in those days of reading and studying at the Pacific Biological Laboratory.

One important evolutionary thinker who emerges early in Ricketts's and Steinbeck's investigations is Jan Christian Smuts. Astro suggests that Steinbeck may first have heard about Smuts from an essay Ritter coauthored with Edna W. Bailey (published in 1931), in which Smuts is praised (*Steinbeck and Ricketts* 48). From an interview with Carol Steinbeck in 1971, Astro confirms that early in the thirties Steinbeck read "Smuts's then-popular essay, *Holism and Evolution*" (48). Joel W. Hedgpeth observes that during the spring of 1932, Smuts's book was one of "Ed's principal household gods" along with W. C. Allee's *Animal Aggregations* and John Elof Boodin's *A Realistic Universe* (*The Outer Shores* 1: 12).

Smuts is a votarist of Darwinian theory; he finds evolution the key to a wider vision, a holistic way of viewing nature, and is even more explicit in linking Darwin to this way of thinking than Ritter. In the preface to *Holism and Evolution,* Smuts writes that "Evolution in the mind of Darwin was, like the Copernican revolution, a new view-point, from which vast masses of biological knowledge already existing fell into new alignments and became the illustration of a great new Principle" (6). In several places in the book he elevates Darwin and evolution, claiming, for instance, that the "view-point of Evolution as creative, of a real progressive creation still going forward in the universe instead of having been completed in the past, of the sum of reality not as constant but as progressively increasing in the course of evolution, is a new departure of the nineteenth century, and it is perhaps one of the most significant departures in the whole range of human thought" (89). Smuts recognized that Darwinian theory presents an examination of the whole and accounts for the interrelations of nature. He appreciated how things taken in this larger view look different—and no doubt come closer to the truth—than things considered in the narrow, pre-Darwinian view of nature as static.

In his enthusiasm for the creative force of evolution, Smuts sees natural selection as a progressive force rather than a destructive force and curiously interprets the struggle for survival as a "form of comradeship, of social co-operation and mutual help" (218). In essence, we are all together in a great whole, and every struggle contributes—through selection—to the progress of the whole. This concept appears in Stein-beck's work as late as *America and Americans*.

Smuts's writing gave Steinbeck a picture of Darwin as a revolutionary thinker who broke through to a way of seeing that perceives an open, natural scheme of progress, interrelation, and cooperation. Steinbeck's fiction in many ways dramatizes such a holistic system. Smuts's work, like Ritter's, is significant because it links Darwin to that system and credits the naturalist's search for truth as largely originating the idea of the whole of nature.

That Steinbeck had read Smuts carefully is evident in *The Grapes of Wrath,* for his definition of a tenant farmer is written in a style that echoes Smuts's definition of a whole. Smuts writes, "A whole, which is *more* than the sum of its parts, has something internal, some inwardness of structure and function, some specific inner relations, some internality of character or nature, which constitutes that *more*" (103). Steinbeck's description of the farmer notes inner relations (chemistry and elements) and external connections with the land, all with the same rhetorical emphasis on "more": "Carbon is not a man, nor salt nor water nor calcium. For he is all these, but he is much more, much more; and the land is so much more than its analysis. The man who is more than his chemistry, walking on the earth, turning his plow point for a stone . . . that man who is more than his elements knows the land that is more than its analysis" (126).

Another of the household gods at the lab was John Elof Boodin; of particular interest here are two books which Steinbeck must have seen, *A Realistic Universe* and *Cosmic Evolution* (DeMott, *Steinbeck's Reading* 16). Steinbeck probably read these immediately after he first heard of Boodin through Albee in the winter of 1932-33 (Benson, *True Adventures* 268). Indeed, of all the naturalist/philosophers discussed here, only Boodin's name appears in *The Log from the Sea of Cortez* (259). As Benson writes, Boodin "was another of the many philosophers of the time who sought to reconcile science, particularly evolutionary theory and scientific methodology, with human values and the concept of an impersonal God as a controlling principle" (267). The idea that "the laws of thought must be the laws of things" (a quote from the introduction to Boodin's *A Realistic Universe* [xvi]) made a lasting impression on Steinbeck; he refers to it not only in *Sea of Cortez* but also in *The Winter of Our Discontent* (DeMott, *Steinbeck's Reading* 135). Boodin's assertion connects the individual mind with other minds and ultimately with the whole of the cosmos.

Cosmic Evolution certainly brought Darwin's theory before Steinbeck's eyes. Darwin's name surfaces again and again in Boodin's book, but this philosopher is not as enthusiastic as Smuts. Boodin does not see natural selection as a creative force, and finds that such a system of chance cannot account for the origin and development of life. He does not discount natural selection, but sees it as only part of the equation. According to Boodin, evolution does not properly consider heredity and mutation as factors. In another partial dissent, Boodin presents Darwinism in a way that actually underscores Smuts's view and have might appealed to the wide-open, nonteleological thinking of Steinbeck and Ricketts: "The modern point of view which finds its typical expression in Darwinism emphasizes change, history, mechanical causes, flux of species. . . . History runs on like an old man's tale without beginning, middle, or end, without any guiding plot. It is infinite and formless" (*Cosmic Evolution* 77). But as he does in other places, Boodin gives Darwin a great deal of credit, at one point connecting the naturalist with an original conception of the whole: "no part can survive which does not enter in some degree into rapport with the whole. Hence the reality of natural selection, the contribution of Darwin" (124). From talk such as this, one might infer that cooperation is necessary for survival—an idea which is crucial in Steinbeck's fiction.

W. C. Allee is the other god in Ricketts's household triumvirate. Ed had taken Allee's course in animal ecology in the fall of 1922 at the University of Chicago (Benson, *True Adventures* 191). Allee's important book, *Animal Aggregations*, was published in 1931, and Ricketts owned a copy (burned in the fire at the lab but replaced by Steinbeck in 1940 [DeMott, *Steinbeck's Reading* 131]). Steinbeck was so impressed with Allee's concept of animal grouping and cooperation that he extended it to human behavior in a "phalanx" theory, which he presented in "Case History" in 1934 and a paper, "Argument of Phalanx," in 1935 (DeMott, *Steinbeck's Reading* 131). The thesis of the phalanx idea is that individuals operate differently apart, but when they key into each other and form a group they become a superorganism with "a will and a direction of its own" (Astro 63). Steinbeck's use of the phalanx theory of human behavior in his fiction is well known, and discussed extensively by Astro, Benson, and Lisca. The theory is best expressed in Steinbeck's short stories, "The Raid" and "The Vigilante" and in the 1936 novel, *In Dubious Battle.*

The theory of evolution is a small factor in *Animal Aggregations,* and Darwin's name never comes up; however, Allee does connect evolution to his principle of cooperation—further evidence of the tremendous effect of evolutionary theory on natural science and the books Steinbeck and Ricketts were reading. In a discussion of animal "survival values," Allee treats natural selection in a way reminiscent of Smuts's interpretation of it as a force of cooperation. Allee writes that "many of the advances in evolution have come about through the selection of co-operating groups rather than through the selection of individuals. This implies that the two great natural principles of struggle for existence and of co-operation are not wholly in opposition, but that each may have reacted upon the other in determining the trend of animal evolution" (361). Allee's point underscores the need for cooperation; animals that cooperate will survive and selfish individuals will perish (a concept dramatized in *The Grapes of Wrath*).

A book that appeared on the shelves of Ricketts's library soon after the fire was Henri Bergson's *Creative Evolution,* which Ricketts purchased in 1937, the year of its publication in America. (DeMott, *Steinbeck's Reading* 13). In his view of evolution, Bergson somewhat follows Boodin's view, arguing that variation and mutation play a much greater role in species change than Darwin believes. One thought that Bergson attributes to Darwin is interesting in light of Steinbeck's discussion of Cathy as a mutation in *East of Eden.* "[Darwin] was not ignorant of the facts of sudden variation," Bergson writes, "but he thought these 'sports,' as he called them, were only monstrosities incapable of perpetuating themselves" (63). In his fiction, Steinbeck seems to have adopted a view closer to Darwin concerning natural selection, for Steinbeck's only portrayal of variation by mutation concerns Cathy, who is indeed a monstrosity.

Two other books that Steinbeck read before 1940 place the human species in the natural scheme, extending the Darwinian view of our subjugation to the same natural laws that affect all animals. In 1933 Steinbeck read Ellsworth Huntington's *Civilization and Climate* (DeMott, *Steinbeck's Reading* 57). The other book, Mark Graubard's *Man the Slave and Master,* was part of Steinbeck's personal library. Stamped "This Book Belongs to Carol and John Steinbeck" (he divorced her in 1942) and printed by Covici-Friede in 1938 (DeMott 47-48), Steinbeck might well have read the book before he wrote *The Grapes of Wrath.*

Huntington attempts to trace the effects of climate on the individual and social human. His book demonstrates that climate has as much bearing on humans as on any other animal and that one of climate's strongest effects upon humans is that it causes "migration, racial mixture, and natural selection" (3). Huntington's observations of how environmental stress leads to migration and the distribution of our species is particularly interesting when applied to *The Grapes of Wrath.* When crops fail due to bad weather, Huntington writes, subsequent economic and political disturbances cause migration, and natural selection in the migrating group follows: "The people who migrate perforce expose themselves to hardships and their numbers diminish until only a selected group of unusually high quality remains" (27). One cannot help but consider the Joads, as migrants, and their rate of attrition.

Before *The Grapes of Wrath,* Steinbeck had already been thinking along these same lines, for he did write of the process of migration and selection by attrition in a series of seven articles for *The San Francisco News* in October 1936 (reprinted in 1988 as a book, *The Harvest Gypsies*). He considers the migrants from the dust bowl as a toughened new breed who, though they have seen many of their numbers die, "have weathered the thing, and they can weather much more for their blood is strong" (*Gypsies* 22). Once in California, the migrants look for land, and the natives must know "that this new race is here to stay and that heed must be taken of it" (22). This dynamic is also considered in *The Log from the Sea of Cortez.*

While the idea of human subjugation to natural laws is evident in *Civilization and Climate,* Graubard makes this point even more explicitly. The object of *Man the Slave and Master,* he writes, is "to present a picture of man's place in the biologic world from the viewpoint of the species as a whole" (i). The book culminates in "A Call For Scientific Humanism," reflecting Graubard's desire to see the true place of the human species by breaking through preconceptions, by adopting a scientifically objective perspective, and by overcoming our selfish anthropocentric point of view.

Like Smuts, Graubard finds the theory of evolution a crucial turning point in human thought: "The concept of evolution established a new methodology and perspective" (157). Although Graubard finds problems in Darwin's ideas, he sees the naturalist's theory as the result of an objective analysis based on extensive observation. Because of Darwin's methods, his was the first "elaborate, lucid and scientific explanation of organic evolution" (160).

Graubard considers Darwin to be one of the men of "great genius" whose work illustrates the kind of scientific method advocated in the latter chapters of *Man the Slave and Master.* In a chapter entitled "The Quest for Knowledge" Graubard outlines this method, in which humans must overcome "selfish interests" that defy objectivity. In his call for scientific humanism, he reiterates the need to sacrifice selfishness and ego: the greed, vanity, and ignorance that are "basically responsible for man's inhumanity to man" (332). A key to humanity is sympathy, "understanding the emotions of others and responding with kindness and compassion" (331). Graubard associates primitive man with this selfishness, vanity, and ignorance and finds that human advancement must include scientific methodology, a wider and less egocentric view, and the overcoming of intolerance with compassion and "scientific humility." The ultimate objective is a society in which all humans get a fair share and no one can be left out, since "scientific humanism takes the welfare of humanity, of the species as a whole, as its keystone" (344). Steinbeck could not have ignored the contribution of Charles Darwin to the biological knowledge of "the species as a whole," a contribution which Graubard acknowledges in his chapter "The Story of Evolution."

Two books that Steinbeck read in preparation for the trip to the Sea of Cortez were William Beebe's *Galapagos, World's End* and Charles Darwin's journal of the voyage of the *Beagle.* Beebe's book, mentioned in the reference section of *Sea of Cortez,* is significant as another source of praise for Darwin that Steinbeck would have seen. Henry Fairfield Osborn's preface cites Darwin as an "immortal naturalist" whose "powers of observation and reasoning were equivalent to a whole previous cycle of human thought" (vii). "Many of Darwin's own views," Osborn writes, "possess this power of inspiration, this inherent quickening force which no passing of years can diminish" (viii). Although focused on his own expedition, Beebe writes of Darwin with a sense of awe; for example, recognizing that Darwin found inspiration for his theory at the Galapagos, Beebe says of James Island: "I realized that I was on classic ground, for Charles Darwin had spent a whole week on shore near this very spot" (150). On the next to the last page, he notes that Darwin's "chapter" on the Galapagos has not been equaled "for general grasp and for sheer interest" (428). Beebe's enthusiasm when writing about Darwin is actually exceeded by Steinbeck's in *The Log*.

Darwin and evolutionary theory hover over the books Steinbeck read at a crucial period in his development as a writer; sometimes Darwin is debated or questioned, but more often he is revered. His name appears again and again and is linked with evolution in a revolutionary perspective, that view of the world that underlines the power of inductive, objective reasoning—a holistic understanding which successfully challenges preconceptions. That linking of Darwin to so much of what Steinbeck admired as well as the fascination the novelist had for evolutionary progress and natural selection help to explain why—of all the writers mentioned above including Allee and Boodin—Darwin's name comes up most often in *The Log from the Sea of Cortez;* in fact, only Boodin's name actually appears in the narrative, a reference lifted from Ricketts's journal of the trip.

PARALLEL EXPEDITIONS:
STEINBECK'S *LOG* AND DARWIN'S JOURNAL OF THE *BEAGLE*

Steinbeck no doubt encountered many of Darwin's ideas beyond evolution through reading and discussions with Ricketts. Smuts summarizes theories in Darwin's The Descent of Man, sexual selection in particular (13, and throughout chapter 8). G. Herbert Fowler's *Science of the Sea*, recommended in the reference section of *Sea of Cortez*, outlines and then debates Darwin's theory concerning the development of coral reefs. Fowler writes that "'Coral Reefs'" is "a book of genius in its line as great as [Darwin's] 'Origin of Species'" (136). Ricketts was even familiar with a theory of Darwin's son, George, considering the effect of tides and moon pull on the development of animals in the littoral (Steinbeck, *Log* 32).

Having encountered so many references to Darwin and his work and considering the intense interest Steinbeck had in biology, he certainly must have encountered *The Origin of Species* at least indirectly and does make reference to it in *Sweet Thursday*. Discussing the "inductive leap" in which "everything falls into place, irrelevancies relate, dissonance becomes harmony, and nonsense wears a crown of meaning," Steinbeck credits Darwin with achieving this "greatest mystery of the human mind" along with such giants as Newton and Einstein (28). "Charles Darwin and his *Origin of Species* flashed complete in one second," Steinbeck writes, "and he spent the rest of his life backing it up" (28). Steinbeck seems to have done a bit of mythmaking here, for certainly Darwin never claimed his theory "flashed complete" in such a way. However, the inductive process the novelist suggests, in which a theory emerges after a gathering of facts, does indicate his knowledge of Darwin's method. Further, Steinbeck is essentially correct in stating that the naturalist spent the rest of his life "backing it up" (Darwin produced six editions of *The Origin of Species* in his lifetime, the last in 1872, ten years before his death).

If indeed Steinbeck had read *The Origin of Species*, he would have seen a perfect illustration of the kind of inductive, wide-open method he and Ricketts so admired. In the first paragraph of *Origin*, Darwin summarizes his inductive process:

> When on board H.M.S. 'Beagle,' as naturalist, I was much struck with certain facts in the distribution of the organic beings inhabiting South America. . . . These facts, as will be seen in the later chapters of this volume, seemed to throw some light on the origin of species. . . . On my return home, it occurred to me, in 1837, that something might perhaps be made out of this question by patiently accumulating and reflecting on all sorts of facts which could possibly have any bearing on it. After five years' work I allowed myself to speculate on the subject, and drew up some short notes; these enlarged in 1844 into a sketch of the conclusions . . . [F]rom that period to the present day [1859] I have steadily pursued the same object (Appleman 35).

Evident in *The Origin of Species* is Darwin's ability to use this method without the interference of ego, preconceptions, or prejudices. As Peter Brent remarks in his biography, *Charles Darwin, A Man of Enlarged Curiosity*, "He could regard the world with an eye that saw only what was in front of it, unobscured by expectation. The only question was whether he had the courage to see with this childlike clarity. It was his lifelong gift and his intellectual salvation that he had" (169). In *The Origin of Species,* Darwin makes this capacity clear when he discovers "after the most deliberate study and dispassionate judgment of which I am capable, that the view which most naturalists until recently entertained, and which I formerly entertained—namely that each species has been independently created—is erroneous" (38). Darwin had the courage to defy universal preconceptions that even he had embraced (before his voyage on the *Beagle,* young Darwin was prepared—albeit half-heartedly—to be a clergyman).

The Origin of Species illustrates the kind of thinking that Ricketts and Steinbeck made their ideal: it shows the attempt to find truth by abandoning popular beliefs, making observations firsthand, gathering the facts together, and achieving the inductive leap to discover a great principle. No wonder Darwin's name appears in the oft quoted definition of nonteleological thinking that Ricketts passed on to Steinbeck for inclusion in *Sea of Cortez:* "Non-teleological ideas derive through 'is' thinking, associated with natural selection as Darwin seems to have understood it. . . . Non-teleological thinking concerns itself primarily not with what should be, or could be, or might be, but rather with what actually 'is'—attempting at most to answer the already sufficiently difficult questions what or how, instead of why" (139).

As an observation of nature from an objective perspective of the whole that is as wide as possible, without preconceptions, *The Origin of Species* provides a great example of "is" thinking. For Steinbeck and Ricketts, nature provides the perfect place to look for "is"; significantly, they share the same view of nature that Darwin not only saw but largely originated: the view of the whole, with humans a part of all natural interrelations. This perspective denies traditionally religious or romantic notions which exalt our species by setting it apart.

Darwin loves the natural world and consistently finds it superior to human artifice. For example, he considers natural selection to be "as immeasurably superior to man's feeble efforts [animal husbandry], as the works of Nature are to those of Art" (Appleman, *Origin* 50). Yet he never romanticizes nature. Because of his recognition of the interrelations and competition among living things, he sees at once both beauty and ugliness: "We behold the face of nature bright with gladness, we often see superabundance of food; we do not see or we forget, that the birds which are idly singing round us mostly live on insects or seeds, and are thus constantly destroying life; or we forget how largely these songsters, or their eggs, or their nestlings, are destroyed by birds and beasts of prey" (50). This realistic view of nature, containing at once its beauty and cruelty, often appears in Steinbeck's fiction; we might recall the savage struggle for survival in the beautiful Torgas Valley of *In Dubious Battle,* or Doc's encounter with the drowned girl in the Great Tide Pool of *Cannery Row.*

Ricketts and Steinbeck would not have missed the holistic perspective evident in the famous "tangled bank" metaphor in the concluding paragraph of *The Origin of Species:* "It is interesting to contemplate a tangled bank, clothed with many plants of many kinds, with birds singing on the bushes, with various insects flitting about, and with worms crawling through the damp earth, and to reflect that these elaborately constructed forms, so different from each other, and dependent upon each other in so complex a manner, have all been produced by laws acting around us" (131). Smuts's book would have pointed this passage out to them, for he quotes it and then writes, "[It is] the expression of a great selfless soul, who sought truth utterly and fearlessly, and was in the end vouchsafed a vision of the truth which perhaps has never been surpassed in its fullness and grandeur" (187). Darwin's metaphor illustrates a perspective that puts every individual organism in sync with all others—nothing, including *Homo sapiens,* can exist outside the biological realities of environment, selection, and evolution. *The Origin of Species,* by its conception and by its subject, illustrates much of what Steinbeck dramatizes in his writing.

For direct evidence of Steinbeck's comprehension of the essentials of evolution—that natural selection derived from the struggle for survival is the primary engine of species evolution—we need look no further than *The Log from the Sea of Cortez.* Steinbeck's fascination for the natural struggle, obvious in many of his entries in *The Log,* significantly does not appear in Ricketts's journal of the trip. (Steinbeck's *Log* is written from what he remembered after the trip, with substantial help from the journal Ricketts wrote as the expedition progressed. This journal was not intended for publication but is printed in *The Outer Shores,* a collection of Ricketts's works edited by Joel W. Hedgpeth.)

Ricketts mentions Darwin only once in his journal, in passing, while discussing dolphins (*Outer Shores* 2: 113). The only other mention of Darwin from Ricketts in *The Log* comes from the essay on nonteleology. All other references to Darwin in *The Log* are from Steinbeck. However, Ricketts did admire Darwin's methodology. In the "Annotated Phyletic Catalogue" at the end of *Sea of Cortez,* the entries on naturalist Philip P. Carpenter do commend Darwin, and these entries were almost certainly written by Ricketts. They are in his style which, as Astro notes, is plagued by clumsiness and breakdowns in syntax (*Steinbeck and Ricketts* 27). Indeed, Steinbeck and Ricketts's editor, Pascal Covici, wanted John to have credit for the narrative in *Sea of Cortez,* and Ed to have credit for the scientific appendices. Steinbeck angrily refused, writing that "this book is the product of the work and thinking of both of us and the setting down of the words is of no importance" (*Steinbeck and Ricketts* 14). Nevertheless, the catalogue's style strongly indicates that Covici had the right idea.

While praising Carpenter's work, Ricketts writes that "like most of Darwin's writings . . . [Carpenter's] transcends its time and subject matter and achieves a quality of universalness" (*Sea of Cortez* 480). In another entry, Ricketts notes that Carpenter's work overcomes the barriers of "indifference, carelessness, and stupidity," like the work of Darwin and a handful of other scientists (482). Clearly, Ricketts admires Darwin as a naturalist and writer.

However, the Darwinian emphasis on survival and species competition in *The Log* is entirely Steinbeck's. Ricketts's journal has a tone of acceptance. His nonteleological consideration of things as they are is perfectly summed up in his comment that "'[g]oing along with' is merely an articulate expression for a process of relaxation whereby you go along with, rather than fight against, the pace of external events over which you have no control" (*Outer Shores* 2: 147). When observing animal life in the Sea of Cortez, Ricketts does not seem interested in the struggle for existence. Steinbeck delights in the great competition.

Peering into the sea near Magdalena Bay, Steinbeck notices the active food chain: "Everything ate everything else with a furious exuberance" (*Log* 48). Nothing like this is found in Ricketts's account. Later, looking into the littoral at Cape San Lucas, Steinbeck observes "a vital competition for existence" where starfish, urchins, and other animals fight back at the pounding sea "with a kind of joyful survival" (*Log* 59, 60). He adds, "This ferocious survival quotient excites us and makes us feel good, and from the crawling, fighting, resisting qualities of the animals, it almost seems that they are excited too" (*Log* 60). This emphasis on survival comes from Steinbeck; the only comment along these lines from Ricketts, observing the same scene, is "it seemed to me that life here is very fierce" (*Outer Shores* 2: 114). The novelist makes a similar note of the competition for existence at Pulmo, also considering the special adaptations made by animals who survive on a coral reef (*Log* 81). Likewise, the struggle for survival in the mangroves near La Paz both fascinates and horrifies him (*Log* 123).

Throughout *The Log*, Steinbeck observes the struggle for survival, and near the end of the book he lets loose his own enthusiasm for what he has seen: "There would seem to be only one commandment for living things: Survive!" In this same passage his understanding of the need for competition becomes clear: "This commandment decrees the death and destruction of myriads of individuals for the survival of the whole" (244). This view is the one that Steinbeck, and not Ricketts, brought to *The Log*.

Steinbeck's fascination with competition often moves to observations of the human species. "It is difficult," he writes, "when watching the little beasts, not to trace human parallels" (*Log* 96). Although he realizes that personification of animals is a pitfall, Steinbeck cannot help but compare our species to others. He notes that a group of humans who dominate an area, driving all competitors out, will actually deteriorate because of the lack of competition. Meanwhile, the surviving remnants of the driven will toughen in their struggle to survive, finally becoming stronger than their atrophied conquerors (96-98) (an idea particularly important in *The Grapes of Wrath*). This discussion reveals Steinbeck's belief that natural selection and evolution can be carried over to *Homo sapiens*. Whether the struggle is among crabs in a tide pool or among people in a lush farm basin, competition weeds out the weak, leaves only the vigorous, and selects the "strong blood" (a term he uses in the articles he wrote about migrant labor in 1936). He speaks of this process again later in *The Log*, commenting on the conquest of the Incas by Spaniards who, eventually atrophied in their conquest, are in turn overtaken by the native Peruvians. He links competition and natural selection to the progress of *any* species: "Perhaps the pattern of struggle is so deeply imprinted in the genes of all life conceived in this benevolently hostile planet that the removal of obstacles automatically atrophies a survival drive" (229).

A very important point, especially if we are to understand the biological slant of Steinbeck's fiction, is that the novelist's interest in Darwin appears to be far greater than that of Ed Ricketts. Comparing Ricketts's journal and Steinbeck's log shows that the emphasis on Darwin and the struggle for survival is the novelist's own. Why? As

Astro observes, Ricketts's approach to life dismisses all "goal-oriented activity" except the concept of breaking through (discovering truth), while Steinbeck "advocates a philosophy of action" (*Steinbeck and Ricketts* 73). In Darwin, Steinbeck saw in many ways the kind of thinker who would agree with his and Ricketts's inductive, holistic method; however, the naturalist's evolutionary theory also presents a scheme of progress, and Steinbeck uses it to dramatize a natural means of human development in much of his writing. Add to this Allee's assertion of cooperation as a key to successful survival, Smuts's view of natural selection as a creative force, anthropologist Robert Briffault's linking of evolution to human social improvement (particularly notable in *The Making of Humanity*, in Steinbeck's personal library [DeMott, *Steinbeck's Reading* 18]), and the result is the process of development for *Homo sapiens* that we find in Steinbeck's work, not Ricketts's, and that originates in Darwinian theory.

Evolution, as a theory based on conflict, naturally lends itself to dramatization in fiction, which feeds on conflict for dramatic tension. We do not have to analyze Steinbeck's fiction very deeply to see that most of his serious works derive from vigorous conflicts, the same kind of fierce struggle for life he so admired in the tide pools.

That Steinbeck understood and adopted the mechanics of Darwinian theory through his reading is certain; just as important, he must have became acquainted with Darwin, the man, through the naturalist's journal of the voyage of the *Beagle*. Steinbeck owned a copy of Darwin's *Journal of Researches into the Natural History and Geology of the Countries Visited during the Voyage of H.M.S. "Beagle"* (DeMott, *Steinbeck's Reading* 32), which, considering the references to the book in *The Log*, he must have read in preparation for the expedition to the gulf. Darwin's book is an account of his voyage as ship's naturalist aboard the H.M.S. *Beagle.* The ship's mission was to survey parts of South America and some Pacific islands while making chronometrical measurements during the entire voyage around the world. Darwin's experiences on the *Beagle*'s five-year mission (from December, 1831 to October, 1836) provided much of the evidence and inspiration necessary for the theory of evolution.

Perhaps somewhat romantically, Steinbeck connects his expedition to the gulf with Darwin's voyage on the *Beagle*. The day the *Western Flyer* returned, putting in at San Diego, Steinbeck was quoted in a United Press release: "[I]t seems that some of the broader, more general aspects of the tie-in of all animal species with one another has been lost since Darwin went out of the picture. We are trying in our small way to get back a phase of that broader view" (Ricketts, *Outer Shores* 2: 2). The quote shows not only that Steinbeck had Darwin on his mind during the trip, but also that he credits the naturalist with a broader view—"the tie-in of all animal species"—an assessment of Darwin which is also emphasized in the introduction to *The Log:* "Our curiosity was not limited, but was as wide and horizonless as that of Darwin or Agassiz or Linnaeus or Pliny" (2). The identification of Darwin as a "wide and horizonless" thinker is one of Steinbeck's highest compliments.

Steinbeck admires Darwin's method and writes wistfully of it in *The Log*. He notes that the mission of the *Western Flyer* parallels that of the *Beagle* as a collecting expedition of species on a sweeping scale, but he laments the acceleration of his trip while envying the pace of Darwin's expedition. Where the *Beagle* could plod on for five years, the expedition to the gulf had to conclude in just six weeks (from March 11 to April 20, 1940). Steinbeck begins a long passage on Darwin's methods by writing that "in a way, ours is the older method, somewhat like that of Darwin on the *Beagle*" (61). Traveling under sail or by horse, Darwin kept "the proper pace for a naturalist" who "must have time to think and to look and to consider" (62).

A key observation follows, which suggests Steinbeck's belief that Darwin's inductive method is indeed the way to capture the whole view that the novelist and Ricketts seek above all: "And the modern process—that of looking quickly at the whole field and then diving down to the particular—was reversed by Darwin. Out of long long consideration of the parts he emerged with a sense of the whole" (*Log* 62). Although Steinbeck concedes that to imitate Darwin's pace, to take to a sailing ship or a horse, would be "romantic and silly," he finishes this passage on Darwin by writing that "we can and do look on the measured, slow-paced accumulation of sight and thought of the Darwins with a nostalgic longing" (62, 63). These comments suggest some firsthand knowledge of *The Origin of Species* since the journal of the *Beagle* chronicles the trip that later inspired the theory of evolution; the "long long consideration of parts" commenced after the ship dropped anchor at Falmouth, England. The journal documents the collecting of those parts from which the whole would emerge, as Steinbeck was keenly aware.

While contrasting in pace, in many ways the expedition of the *Western Flyer* does parallel that of the *Beagle*. Both are journeys of collection and observation, albeit on colossally different scales, narrated by men uniquely humble and uniquely bent on recording truth.

Darwin's journal is an account of his observations, and one of the many themes that emerges concerns the superiority of firsthand experience over secondhand authority. For example, Darwin counters the Spanish geographer Felix de Azara's comments about the agouti (a harelike animal of South America) with observations of his own (Browne and Neve, *Beagle* 86-87). From his record of animal populations near Cape Town, he likewise overturns the notion that large animals require lush vegetation, a "prejudice [which] has probably been derived from India" (98-99). Darwin also finds from experience that authorities on the Tahitians have misled him and dismisses the belief that these Polynesians have become a gloomy people (301). After an arduous trip up the Santa Cruz River in Patagonia, the naturalist wryly comments on the myth that going a little hungry can be a good thing: "Let those alone who have never tried it, exclaim about the comfort of a light stomach and an easy digestion" (170).

Pitting observation against authority, Darwin delights in any discovery of truth that overturns an accepted falsehood. He augments this tendency with occasional comments upon the folly of ill-conceived theorizing. Encountering the half-buried remains of a mastodon, he is amused at the locals' deduction concerning the fossil: "the necessity of a theory being felt, they came to the conclusion, that . . . the mastodon formerly was a burrowing animal!" (124). He also dismisses the supernatural explanation given by the inhabitants of Concepcion, Chile, for the eruptions of a nearby volcano as a "silly belief" borne of a need "to apply the witchcraft to the point where their knowledge stopped" (233). In the conclusion of his journal, Darwin warns against making deductions without the proper observations: "I have found to my cost, a constant tendency to fill up the wide gaps of knowledge, by inaccurate and superficial hypotheses" (377).

The Log from the Sea of Cortez also is a journal of observations, and Steinbeck writes in his introduction that the goal of the trip, "to observe the distribution of invertebrates," justifies calling it an expedition (1). The plan was to go wide-open, as Steinbeck understood Darwin to do, and "see what we see, record what we find, and not fool ourselves with conventional scientific strictures" (3), blinders that shut out a view of the whole. In the passage concerning Darwin's method, Steinbeck comments on the way of "our time" in which "the young biologists [are] tearing off pieces of their subject, tatters of the life forms, like sharks tearing out hunks of a dead horse, looking at them, tossing them away" (62).

As a book with its basis in the observed world, *The Log*, like Darwin's journal, criticizes false deductions. In a passage reminiscent of Darwin, Steinbeck discusses facts that contradict a theory explaining the similarity among flora and fauna of the west and east coasts of Baja. From this discussion, he writes of the scientific method in general, noticing the tendency to preserve a cherished hypothesis that has become "a thing in itself, a work of art" despite facts that shoot holes in it (182). He cites the examples of a "leading scientist" trying to find a reef, despite no indicative soundings, because "his mind told him [it] was there" and the expedition of the members of a "learned institution" who could not accept the existence of sea otters once they had erroneously determined that the animals were extinct (62, 183). As Darwin would find when he advanced his theory of evolution on a creationist world, Steinbeck notes that "beliefs persist long after their factual bases have been removed" (183).

The two journals represent attempts to record the world in the most true way; these are books in which factual observations reign supreme, where no quarter is given to error. Yet both authors, meticulous in the desire to be honest, acknowledge the difficulty of being true and objective.

Darwin not only recognizes his ignorance or inadequacies throughout the journal, he also occasionally comments upon the warp of his perceptions. After leaving the confines of the *Beagle* (where he suffered horrible bouts of seasickness), the naturalist knows that he sees the "very uninteresting" country of Maldonado in South America with heightened affection (Browne and Neve, *Beagle* 71). He notices how the experience of a severe earthquake in Chile destroys one's belief in the solidity of earth, therefore conveying to the mind "a strange idea of insecurity" (229). Darwin also comments on the inadequacy of language or preserved specimens to convey the reality of his experience. For example, he discusses how exotic butterflies, preserved in the entomologist's dark cabinet, lose the grandeur of flight observed under the sunshine of the tropics. After describing San Salvador, he finally admits that "to paint the effect is a hopeless endeavour" (367). He writes that "one wishes to find language to express one's ideas" but that "epithet after epithet is found too weak to convey . . . the sensation of delight which the mind experiences" (367).

Steinbeck likewise recognizes that *The Log*, though an attempt to observe invertebrate distribution in a most objective way, will nevertheless be "warped, as all knowledge patterns are warped, first, by the collective pressure and stream of our time and race, second by the thrust of our individual personalities" (2). In a vein similar to Darwin's comments on the preservation of butterflies, Steinbeck writes that one can pickle a Mexican sierra (a type of mackerel) for species identification, but the experience of the living sierra in its natural habitat will be lost. Despite preserved specimens and accurate descriptions, the full sense of the Sea of Cortez must elude the reader—in short, no objective, absolutely clear presentation is possible. Because Steinbeck and his companions enter the gulf, giving and taking from it, they can never "observe a completely objective Sea of Cortez anyway" and, therefore, will not "be betrayed by this myth of permanent objective reality" (3). Although striving to observe and document the truth, neither Steinbeck nor Darwin has an ego big enough to believe that he will record the perfect picture.

Yet Steinbeck and Darwin certainly found their own version of "what actually 'is'": a view of the whole. In his journal, Darwin comments that full comprehension of the beauty of a particular land is enhanced by knowledge of the whole world—exotic lands and one's familiar home both appreciate in beauty by their comparison. Darwin writes that "as in music, the person who understands every note will, if he also possesses a proper taste, more thoroughly enjoy the whole, so he who examines each part of a fine view, may also thoroughly comprehend the full and combined effect" (Browne and Neve, *Beagle* 373). When we consider the whole in regard to the natural scheme, we can realize our place in the world. As Steinbeck recognized, Darwin's conception of the whole of nature—his wide view—is what made the theory of evolution possible.

Of course, Steinbeck considers the whole in *The Log*, and like Darwin he considers things as they are, in their totality and interrelations. Steinbeck writes, "The whole picture is portrayed by *is*" (154). An excellent example of this kind of thinking occurs in the discussion of Japanese fishing boats in the gulf, in which the problem of ecological exploitation is seen from a holistic perspective (264-65). These dredge boats scooped up the bottom for shrimp and killed anything else accidentally caught. Expanding upon comments found in Ricketts's journal, Steinbeck observes that the incredible slaughter represents a terrible waste, particularly to the Mexican people who sorely need the squandered resource. Yet Steinbeck likes the Japanese fishermen, realizing that from their viewpoint they have no idea of what they are doing. Later, after a fisherman crewing on the *Western Flyer* complains of the waste, Steinbeck widens the view, recalling that although humans will suffer due to the loss, other creatures—from the sea gulls gorging what they can at the surface to bacteria feeding on detritus at the sea bottom—will benefit from the slaughter. This is the whole perspective at work: "To each group, of course, there must be waste—the dead fish to man, the broken pieces to gulls, the bones to some and the scales to others—but to the whole, there is no waste" (265).

This view of the ecosystem shows the human as just another link in the food chain. Such thinking slips outside the limited wants of *Homo sapiens*. Steinbeck writes that our "most prized" understanding is "the attempt to say that man is related to the whole thing," a viewpoint that "made a Jesus, a St. Augustine, a St. Francis, a Roger Bacon, a Charles Darwin, and an Einstein" (218). So often when Steinbeck speaks glowingly of the whole viewpoint, the recognition of "is," Darwin's name appears.

Because of the relative similarity of the expeditions, at least in respect to naturalism, coupled with the kindred attitudes of the authors (Darwin just discovering his inductive methodology and Steinbeck in full comprehension of it thanks to his reading and the influence of Ed Ricketts), both journals have the same point of view. In his comparisons of the expeditions of the *Western Flyer* and the *Beagle*, Steinbeck recognizes the similarities in approach as well as the differences in scope and *Zeitgeist*. An understanding of Darwin's methods and theories, as well as his contribution to holistic thought, would have been enough to impress an amateur biologist like Steinbeck, but his reading of the *Beagle* journal acquainted him with another dimension: Darwin's personality.

The tone of the following passage from *The Log* echoes the same praise for Darwin found in the books that Steinbeck read; no one else cited in *The Log* is written about with such unabashed enthusiasm:

On this day, the sun glowing on the morning beach made us feel good. It reminded us of Charles Darwin, who arrived late at night on the Beagle in the Bay of Valparaiso. In the morning he awakened and looked ashore and he felt so well that he wrote, "When morning came everything appeared delightful. After Tierra del Fuego, the climate felt quite delicious, the atmosphere so dry and the heavens so clear and blue with the sun shining brightly, that all nature seemed sparkling with life." Darwin was not saying how it was with Valparaiso, but rather how it was with him. Being a naturalist, he said, "All nature seemed sparkling with life," but actually it was he who was sparkling. He felt so very fine that he can, in these charged though general adjectives, translate his ecstasy over a hundred years to us. And we can feel how he stretched his muscles in the morning air and perhaps took off his hat—we hope a bowler—and tossed it and caught it.

On this morning, we felt the same way at Conception Bay. "Everything appeared delightful." (195-6)

Of all the biologists and scientists that Steinbeck read, only Darwin inspired such personal affection. Here, in the work in which Steinbeck discusses more than anywhere else his biological philosophy, the novelist's identification with Darwin is presented with longing. Inspired by the passage in Darwin's *Beagle* journal, Steinbeck not only wants to bridge a century by reading—he does so by imagining as well. His mental picture of Darwin "sparkling with life" and tossing a bowler is not only enthusiastic but rosy. He paints Darwin in the same tone as he paints Ed Ricketts; earlier in *The Log*, he pictures Darwin scooping up a jellyfish or staring into the sea over the ship's rail. Steinbeck praises this man whose thinking and writing possess "the slow heave of a sailing ship, and the patience of waiting for a tide" (*The Log* 62). Like his methodology, Darwin's personality also translates over a hundred years to Steinbeck.

The man whom Steinbeck saw portrayed in the journal of the *Beagle* is one of humility, fairness, and compassion. Darwin readily admits his own shortcomings ("I am quite ignorant" appears twice in the first chapter [Browne and Neve, *Beagle* 52, 53]). Indeed, Darwin's biographers, from his son Francis to Peter Brent, note the humility of his personality. Like Steinbeck, Darwin did not feel comfortable with fame—he escaped accolades and criticism after the publication of *The Origin of Species* in 1859 by secluding himself at his house at Down, his refuge from 1842 to the end of his life. Steinbeck ducked the spotlight by going to the gulf amid the popularity of *The Grapes of Wrath*. Several biographers, such as Brent, and R. Colp Jr. in his *To Be an Invalid: The Illness of Charles Darwin*, have suggested that Darwin's prolonged illnesses—particularly the headaches—were stress related, much the same as Steinbeck, whose illness, depression, and general misery after the uproar caused by the publication of *The Grapes of Wrath* are chronicled by Benson in *The True Adventures of John Steinbeck, Writer* (and by Steinbeck himself in *Working Days*, the journal he wrote while creating the novel). Perhaps Steinbeck, reading the generally humble tone of Darwin's words, recognized a kindred spirit.

Probably most endearing to Steinbeck would have been Darwin's obvious compassion and empathy for the underdog. During his travels, Darwin demonstrates a dislike for slavery, for penal colonies, and for the exploitation of workers by landowners. While in Brazil, he characterizes the threatened separation of the members of a slave family as "one of those atrocious acts, which can only take place in a slave country" (Browne and Neve, *Beagle* 62). In Australia, he notes a serious drawback for wealthy families served by convicts: "How thoroughly odious to every feeling to be waited on by a man, who the day before, perhaps, was flogged, from your representation, for some trifling misdemeanor" (326). In a footnote on the condition of agricultural workers in Chile, he recognizes the extreme poverty of the laboring class and writes that the miners in Chile are treated like horses by their masters (214). On the island of Terceira he observes that the handsome and good-humored peasantry emigrate to Brazil to escape poverty, only to find that "the contract to which they are bound differs but little from slavery" (370). Steinbeck, writing *The Grapes of Wrath* from his observations of the degradation of California laborers, would surely have found Darwin's accounts of human exploitation compelling.

The novelist might also have been pleased to learn of Darwin's appreciation for the primitive life. The naturalist is quite taken with the gauchos he traveled with in Argentina and Patagonia, even though he knows they can be either extremely polite or "seem quite as ready, if occasion offered, to cut your throat" (74). He writes of their free, independent lifestyle with admiration and a clear, simple, poetic style similar to Steinbeck's: "There is high enjoyment in the independence of the Gaucho life—to be able at any moment to pull up your horse, and say, 'Here we will pass the night.' The deathlike stillness of the plain, the dogs keeping watch, the gipsy-group of Gauchos making their beds round the fire, have left in my mind a strongly marked picture of this first night, which will not soon be forgotten" (85-86). In a similar passage, Darwin describes a gaucho urging a horse to swim across a river: "A naked man, on a naked horse, is a fine spectacle; I had no idea how well the two animals suited each other" (139). Considering Steinbeck's consistent dislike for the "respectable" class, he would have enjoyed Darwin's opinion that while some of the noble traits of the gaucho may be found in the "higher and more educated classes" of the country, these worthies are "stained by many vices of which he [the gaucho] is free" (144). Darwin's occasional disparaging observations of the leaders in South America, such as General Rosas (a man bent on the inhuman slaughter of native peoples), would no doubt also have been well received by Steinbeck.

Steinbeck's general admiration for primitive living and his suspicion of the elite classes is explicit in *The Log*. Speculating on the destruction of the civilized human races by the "mutation" resulting from the industrial revolution, Steinbeck believes that the so-called "degenerates"— the Native Americans of Lower California—will survive after the "godlike race" flies away in bombers "to the accompaniment of exploding bombs" (91). Later, speaking again of Native Americans, Steinbeck compares them favorably to the supposedly civilized world. He does not want to write of them as noble savages, and he acknowledges the shortcomings of their lives, yet he admires their simplicity as Darwin does the gauchos'. After a passage describing many of the ills of modern, mechanized society, Steinbeck writes that "to an ignorant Indian these might not be evidences of a great civilization, but rather of inconceivable nonsense" (210). Like Darwin, Steinbeck analyzes the "primitives" with a desire not to judge them as simply backward or inferior.

In their writing, both Steinbeck and Darwin demonstrate a strong compassion for those oppressed, sharing a tone of selflessness and sympathy. At the end of his journal, Darwin recommends to future naturalists who go abroad that they possess "good-humoured patience" and "freedom from selfishness" (Browne and Neve, *Beagle* 377). Steinbeck attaches selfishness to the "intellectual" mule who "dwells inward in sneering intellectuality" and is "sullen, treacherous, loving no one, selfish and self-centered" (*Log* 164). In the next paragraph, Steinbeck defines *"simpatica"* as not simply passive sympathy but a form of active cooperation (165).

To be sure, Darwin does at times exude an air of pride for Victorian England and his country's superiority (as Steinbeck does, for example, in his World War II propaganda piece, *Bombs Away*). The naturalist can be hard in his judgments—particularly concerning the indigenous people of Tierra del Fuego—but overall his journal presents a man of openness and sympathy. Considering that the method and point of view of Darwin's journal parallel *The Log* in many ways, we need not wonder that Steinbeck enthusiastically compares his expedition to Darwin's. The naturalist pioneers a way of thinking that the novelist not only adopts, but spends most of a lifetime depicting.

To some extent, the parallel expeditions of Darwin and Steinbeck leave even Ricketts behind. Darwin translates over a hundred years to Steinbeck, indirectly through a range of writers like Ritter, Smuts, Boodin, and Allee and directly through the *Beagle* journal and probably also *The Origin of Species*. Steinbeck seized on the Darwinian concepts of competition and natural selection. The expeditions are parallel, as journeys of collection and observation and as inductive approaches to the whole. In point of view, the journal and *The Log* are similar as presentations of honest, open, and compassionate writers. Indeed, if we use Darwin as a pilot, an expedition through many of John Steinbeck's books reveals a unique, cohesive approach to fiction, to seeing, to truth, and to life. The rest of my study will be a Darwinian reading of Steinbeck's work, a look at that biological view of our species that the novelist has so carefully charted.

3

Beasts at the Door

"The vast mass of mankind remains to-day, under the external appearance of transforming civilization, at heart much what they have been in the rudest ages, barbarians . . . and liable to break out at any time through all their veneers into primitive savagery and barbarism."
—*Robert Briffault,* The Making of Humanity, *1919*

Much of Steinbeck's fiction challenges our notions of the exalted human position, reminding us that we answer to many of the same natural laws and conditions that the occupants of the tide pool do. As a more sophisticated species than a crab or anenome, our struggles for dominance and survival may be complicated and veneered by civilization, but often enough the human competition is so fierce that the tide pool seems like a playground. This hard look at what it is to be human underscores the part of Steinbeck's work that, far from sentimental, is quite brutal. Steinbeck's emphasis on competition and his attack upon the veneer of civilization demonstrate, like few other authors, how thin the line is between human and beast.

Steinbeck, like Darwin in *The Descent of Man,* sees the human being as just another species in a world of species. To take this view, one must break through many preconceptions—both humanist and religious—in order to deny anthropocentric renderings of the ecosystem. This is a very difficult position, likely to attract criticism from others unwilling to go so far. Critics from Wilson to Fiedler cannot accept Steinbeck's view, just as creationists today have made a science out of denying Darwin's theory of evolution. Yet Darwin and Steinbeck both knew that if we are to get to the truth about our place in the world, we must recognize ourselves as animals connected inextricably to a larger natural whole, a viewpoint underscored today by the environmental movement.

At the end of *Descent*, Darwin reminds us of our connection to the other animals and our own ancient past, writing that "with his god-like intellect which has penetrated into the movements and constitution of the solar system—with all these exalted powers—Man still bears in his bodily frame the indelible stamp of his lowly origin" (Appleman 208). In *The Log from the Sea of Cortez*, Steinbeck reaches a similar conclusion, but asserts his view more aggressively than Darwin. "Why do we so dread to think of our species as a species?" Steinbeck asks. "Can it be that we are afraid of what we might find? That human self-love would suffer too much and that the image of God might prove to be a mask?" (266). Steinbeck goes on to suggest that we live in self-delusion, that "if we could cease to wear the image of a kindly, bearded, interstellar dictator, we might find ourselves true images of his kingdom, our eyes the nebulae, and universes in our cells" (266). Steinbeck recognizes that our inability to see ourselves as a *species*, living in the preconception that we are in the image of a god, denies us a true picture, a better perception of ourselves which would connect us to the nebulae and the microcosms in our cells—the whole.

These passages help to explain Darwin's eminence in *The Log*. The theory of evolution, as so much of Steinbeck's scientific reading indicates, makes the novelist's perception of *Homo sapiens* as a species possible. From *The Origin of Species* Darwin had begun a revolutionary line of thinking that he bravely pursued; evolution, if a true theory of natural species development, could be applied to humans. "As soon as I had become, in the year 1837 or 1838, convinced that species were mutable productions," Darwin writes, "I could not avoid the belief that man must come under the same law" (*Autobiography* 49). *The Descent of Man*, first published in 1871, is Darwin's discussion of human development as a species. Although we have no evidence that Steinbeck actually read Darwin's *Descent* (some of Darwin's theories, however, particularly sexual selection, are summarized in Smuts's *Holism and Evolution*), what is interesting is the evidence from *Descent* that Darwin's conception of *Homo sapiens* as a species parallels Steinbeck's. Once Darwin and Steinbeck had adopted the view of the natural whole, noting the important interrelations of species and the effect of natural principles upon all life, both men considered the human being as part of that whole. To do so was a scientific revolution for Darwin and a literary revolution for Steinbeck.

Concluding *The Descent of Man,* Darwin predicts that the main idea of the book, "that man is descended from some lowly organized form, will, I regret to think, be highly distasteful to many" (Appleman 919). From his long experience with the reaction to *The Origin of Species,* he knew that the extension of the theory of evolution to human beings would conflict with many people's theological and scientific beliefs. Yet he wrote the book, a brave treatise that portrays us as subject to natural and sexual selection. In the introduction, he indicates that *The Descent of Man* will place humans side by side with other animals, as it will consider "whether man, like every other species, is descended from some pre-existing form" (390). Darwin spends much of the first half of his book observing physiological and mental parallels between humans and other animals, particularly the primates. He shows how we are connected to other animals by similarities in physical form, including the structure of the brain and "all other parts" of the body. We bear the same hardships of the mammals about us, suffer the same parasites, and often heal in the same way. Considering the effects of lunar cycles, he writes: "Man is subject, like other mammals, birds, and even insects, to that mysterious law, which causes certain normal processes, such as gestation, as well as the maturation and duration of various diseases, to follow lunar periods" (397). Ricketts and Steinbeck believed in this concept as well, and the novelist put it to use in his unpublished novel, "Murder at Full Moon."

Darwin asserts that we are so close in development to other primates that we cannot claim a distinct order for ourselves in the animal kingdom (Appleman, *Descent* 517), even joking in a provoking way about just how close we are to them: "An American monkey, an Ateles, after getting drunk on brandy, would never touch it again, and thus was wiser than many men" (397). In a more serious vein, Darwin concludes in *Descent* that he would rather be like a monkey he heard had risked its life to save its master than be like the depraved "savages" he witnessed in Tierra del Fuego (919-920). Although Darwin certainly recognizes humanity's singular mental achievements (such as the sophistication of symbolic language), his book studies our true place in the natural world. The stamp of the past is upon our brow; we are a *species,* however closely or distantly related to all other animals.

Whether or not he had read *The Descent of Man,* Steinbeck largely shared the view of this work—his holistic and biological perspective would naturally lead him to the same conclusions. Crucial to Steinbeck's vision is his Darwinian placement of *Homo sapiens* in the whole. Almost all of Steinbeck's writing presses against the fragile and often illusory line between what is human and what is animal, affecting its style and themes. Steinbeck's fascination with the struggle for survival strips off the veneer of civilization; for example, a novel like *In Dubious Battle* is less concerned with thirties labor politics and more concerned with showing, in a violent struggle for dominance, how the old law of the tide pool emerges, and hence the darker side of Steinbeck's dramatization of Darwinian theory.

THE LANGUAGE OF THE SPECIES

"Perhaps we will have to inspect mankind as a species," Steinbeck writes in his last lengthy publication, *America and Americans*, "not with our usual awe at how wonderful we are but with the cool and neutral attitude we reserve for all things save ourselves" (137). Late in his life, Steinbeck still had not lost this "neutral attitude." Indeed, in an unpublished 1993 interview, Elaine Steinbeck debates the popular critical notion that her husband dropped his scientific ideas and somehow changed completely after the forties. She notes that he did not lose his interest in marine biology and bought a house in Sag Harbor because he wanted to be near fishermen and people who worked by the sea. "When we went to Nantucket, he always went to Woods Hole," she explains. "When we went to the Caribbean [beginning in 1952] . . . we stayed at St. John. There was a marine biology station on that island and he always went over. . . . He never gave up his interest in the sea or marine biology." She adds that one of his great disappointments was that he never could visit the Great Barrier Reef. Although she does not recall specific conversations with Steinbeck about Darwin, she is certain that the naturalist "influenced John tremendously and continued to all the rest of his life."

Throughout much of Steinbeck's work, he tries to stand outside humanity and examine it as a biologist would any other species. This perspective accounts for the relative rarity of the first person point of view in his fiction; even in the nonfiction essays, including the last long one, *America and Americans,* he restrains his use of "I." To step outside, maintaining the neutral perspective, is Steinbeck's favorite mode of approaching the truth—to coolly observe the species helps to avoid those illusions and preconceptions that tell us "how wonderful we are." The species view becomes implicit in his narration, demonstrated again and again stylistically by his paralleling of humans and animals.

Steinbeck's use of such parallels has been considered in various ways by critics, beginning with Edmund Wilson's disdain for the novelist's apparent preoccupation with animalism (48). Discussing *To a God Unknown,* Peter Lisca notes Steinbeck's parallel of human and animal actions. He writes that Steinbeck's narrative may be interrupted by a predatory incident in nature that has some relation to the story; such incidents "throw light on the moral structure of that ultimate reality with which man is consanguineous—Nature" (*Wide World* 54). F. W. Watt recognizes Steinbeck's analogy of human activity and the tide pool, where dramas of survival and society appear in miniature (16-17), and Stanley Alexander notices the same analogy at work in *Cannery Row.* Astro considers Ed Ricketts's interest in human/animal parallels and his view of tide pools as "replicas" of human social orders (*Steinbeck and Ricketts* 40). A note Steinbeck penned while preparing to write *In Dubious Battle* suggests he would agree with Watt, Alexander, and Astro: "Come down to the tide pool, when the sea is out and let us look into our old houses, let us avoid our old enemies." Of course, Steinbeck's most famous parallels, that of the wandering turtle and the migrant workers in *The Grapes of Wrath,* and the transformation of Pepé into an animal in "Flight," have been discussed by numerous critics. What is necessary, however, if we are to understand the power of his holistic, Darwinian view of *Homo sapiens,* is to consider thoroughly Steinbeck's employment of parallels as they shape his style as a writer.

One of the interesting characteristics of Steinbeck's prose is that he often uses the word "species" rather than "mankind" or "humanity." Talk of species and strong blood enters Steinbeck's consideration of Depression migrations in *The Harvest Gypsies* and *The Grapes of Wrath.* We find it in *Sweet Thursday,* also, in the gruff assessments of Old Jingleballicks (184), or when Doc confesses he has thought of Suzy as he would a species of octopus (187). Joe Saul in *Burning Bright* can finally accept the son he did not father when he realizes the big picture: that it's not his line that matters so much as it is the progression of the human species (105). And in the opening of *The Wayward Bus,* the lunchroom calendar girls and the silver coffee urn are considered from the point of view of another species looking at our own (3, 5).

Often, Steinbeck's observations of the species coincide with the topic of survival. In the introduction to *Once There Was a War*, he writes that if we, a "species" prone to the "accident" of war, are stupid enough to have a third global conflict, "we do not, in a biologic sense, deserve survival" (v,vi). He adds bitterly, "There is no reason to suppose that we are immune from the immutable law of nature which says that over-armament, over-ornamentation, and, in most cases, over-integration are symptoms of coming extinction" (vi). Here his view that we, as a species, are subject to the "immutable law of nature" is explicit. Ironically, in the propaganda piece *Bombs Away* (a book on the training of bomber crews written for the government in 1942), he suggests that war has a positive effect as its threat rouses America to action: "The attack on us set in motion the most powerful species drive we know—that of survival" (14).

In *Travels with Charley*, Steinbeck considers Americans in terms of natural selection and survival. He writes of the restless breed of people who came to America and their impact on the development of the country. Taking the wider view of *Homo sapiens* in general, he concludes that "we are a restless species with a very short history of roots, and those not widely distributed" (104). Discussing progress and its ecological impact, he realizes that we feel not only the constraints of poor weather or plague: "Now the pressure comes from our biologic success as a species" for "we have overcome all enemies but ourselves" (196). Looking at the Mojave desert, which he writes has been a testing ground to see if all comers are "good enough to get to California" (209), Steinbeck summarizes the processes of natural selection, demonstrating unequivocally the pervasiveness of Darwin's concept in his thinking:

> One ingredient, perhaps the most important of all, is planted in every life form—the factor of survival. No living thing is without it, nor could life exist without this magic formula. Of course, each form developed its own machinery for survival, and some failed and disappeared while others peopled the earth. . . . If the most versatile of living forms, the human, now fights for survival as it always has, it can eliminate not only itself but all other life. And if that should transpire, unwanted places like the desert might be the harsh mother of repopulation. For the inhabitants of the desert are well trained and well armed against desolation. (215-216)

Here, in a veiled discussion of nuclear war, Steinbeck dwells less on the political sphere and far more on the biological picture. In the great scheme of natural selection, the human has proved a survivor, but is perhaps doomed to self-destruction. We may create a desert of our Earth, and only those creatures used to a wasteland will be selected out and thrive while the rest disappear. He concludes this discussion with the image of a lone human couple who might emerge "with their brothers in arms" (216), a host of desert animals from armored insects to coyotes. This is a typical Steinbeck picture: humans linked with other animals, "brothers," struggling to survive under the same natural laws.

Studying the good and bad of American society, *America and Americans* clearly looks at its subject, playfully named *Genus americanus*, as a species. Steinbeck charges that Americans have been glutted by wealth and are dulled to natural instincts such as the survival drive because they have become atrophied by comfort, an opinion reminiscent of his discussion about the conquering Spaniards or the lazy *botete* fish in *The Log* (229, 128). In the final chapter, in a way Steinbeck's swan song, he confronts this problem from the scientific perspective he is most comfortable with: "[W]e must inspect the disease as a whole because if we cannot root it out we have little chance of survival" (137). To isolate the "disease," he calls upon us to look at ourselves as a species.

Steinbeck sees our problems, from crime to racism, in biological terms; we are like a kennel of pedigreed bird dogs sitting about unused. "For a million years we had a purpose—simple survival," he writes. "This was a strong incentive" (139). Now, he asserts, the American species suffers from the "terrible hazard of leisure" (139). Our numbers increase while, without a need for the basic drive of survival, our purpose disintegrates. Angrily, Steinbeck writes that "we give lip service to survival"; "neither the sleeping pill, the church, nor the psychiatrist" can hide the fact that too many people are born and grow up with little to do (141). Without the urgency of the struggle for existence, life cheapens. What would seem an ironic statement, but extends from Steinbeck's Darwinian conception of struggle, selection, and evolution, is the author's belief that the only hope for Americans is their sense of discontent. "We have never sat still for long," he concludes. "[W]e have never been content with a place, a building—or with ourselves" (143).

Explicit in Steinbeck's nonfictional essays is his biological view of people as a species. Forgetting that biological vision will lead to a befuddled view of his art. His depictions of strikes are anchored in something deeper than thirties socialism; his Christ figures (like Casy) or angelic personae (like the Seer of *Sweet Thursday*) will not fit any theology. The species view is implicit in the narration of his novels. What Darwin aims to do in his theories, to connect humans to the whole, Steinbeck achieves through setting, personification, anthropomorphism, theriomorphosis (rendering humans with animal characteristics), animal/human parallels in action, and placement of humans and animals in close quarters.

Setting

Perhaps reinforced by his reading of Ellsworth Huntington's *Civilization and Climate,* Steinbeck connects humans to their environment by linking climate and geography to the moods of his characters. This handling of atmosphere is not a simple case of naturalism, for the environment does not merely shape characters—sometimes the perceptions of the characters shape the land or ecologically alter it. His view *appears* to be a romantic one, particularly the idealized interactions of human and nature that Ralph Waldo Emerson describes in *Nature.* R. W. B. Lewis writes that the "sky-blue vision of things" in *The Grapes of Wrath* is "straight out of Emerson," and therefore accuses Steinbeck of romantic writing that clashes with the harsh realities of depression-era California (173). However, Steinbeck himself pointed out in a 1954 letter to James D. Brasch: "I am afraid you will have to set the Emerson likeness down to a parallelism rather than a direct influence." In fact, the confluence of human and place stems from his scientific viewpoint.

Steinbeck's emphasis on setting, necessary for his biological perspective, has attracted varying critical commentary. Unfortunately, it has helped pigeonhole him as a regionalist. Noticing the importance of setting in Steinbeck's fiction, particularly the Salinas Valley, F. W. Watt writes that "despite the considerable variety in his writing Steinbeck has remained at his best a regional novelist" (3). Warren French, although not labeling Steinbeck a regionalist, suggests that when Steinbeck left his native California, he lost a sense of place in such novels as *East of Eden* and *Sweet Thursday* (162). Comparing Steinbeck's use of place rather unfavorably to Faulkner's, John Ditsky identifies the California novelist's artistic struggle as "trying repeatedly, with varying degrees of success, for a restrained imaginative employment of his often unruly California species of the otherwise universal truths of Nature" ("Faulkner Land and Steinbeck Country" 12). However, breaking out of regionalist typecasting, Ditsky asserts that one of Steinbeck's most successful uses of place is with the New England setting of *The Winter of Our Discontent* (22).

Geography serves as a unifying element in Steinbeck's fiction, as several critics have observed of *Pastures of Heaven,* and Arnold Goldsmith has noted concerning *The Red Pony.* R. S. Hughes applauds the way in which Steinbeck's stories "evince a unique spirit of place," particularly the California coast and inland valleys (118), and Louis Owens finds place to be so important in Steinbeck's works that the critic categorizes them by setting and writes that "America has produced no writer more acutely conscious of setting than Steinbeck" (122). Owens's judgment may be true, for the dramatic effect of climate and geography upon characters is too frequent in Steinbeck's fiction to document here. However, looking at some examples throughout his career will help demonstrate how setting ties into his holistic conception of human and nature, and should help to explain why setting is so important in his work.

Cup of Gold (1929), Steinbeck's most romantic work, was written as a passionate swashbuckling tale, a fictionalized account of the infamous seventeenth-century pirate, Henry Morgan. Still, we find examples of climate paralleling the actions or moods of characters. The night that Dafydd, a haunted sea dog reminiscent of the ancient mariner, returns from the sea to visit the Morgan home is one of "preternatural" mystery, with crying winds, inspiring young Henry to look beyond the walls of the house for "unbodied things" (4). These few parallels do seem muddled in the pathetic fallacy of romanticism, but, as Moore has noticed, very quickly in Steinbeck's career such images become more rooted in science. Earlier unpublished stories that were not written in the shadow of the swashbuckler tradition make use of setting that, while still romantic, seems inspired by the holism of Steinbeck's Stanford influences, Ritter and Brown. These settings emphasize natural imagery, even in such unlikely places as Los Angeles or New York.

It should be noted that some controversy surrounds the authorship of four of these early stories: "The Days of Long Marsh," "East Third Street," "The Nail," and "The Nymph and Isobel." The stories, surviving as typed, unsigned carbons at Harvard's Houghton Library, are catalogued there as "Formerly incorrectly attributed to John Steinbeck." Elizabeth R. Otis, his agent and representative of his estate, informed the editor of the *Steinbeck Quarterly* in 1971 that the stories "had been erroneously attributed to Steinbeck [and] are *not* written by Steinbeck" (note 62). However, Benson demonstrates that these stories were Steinbeck's by reviewing the author's 1924 letters to a friend, Carl Wilhelmson, where references are made to the stories (*True Adventures* 75-76). In tone and style, they bear the unmistakable stamp of John Steinbeck, "The Days of Long Marsh" in particular. As John Timmerman noted in 1990, the stories "are now generally acknowledged to be Steinbeck's" (*Dramatic Landscape* 22). One likely explanation for the question of the stories' authorship is in another catalog note, apparently with them from the time they came to Harvard through the Byerly fund on October 15, 1941: "not to be published during the author's lifetime." The note would certainly be in line with Steinbeck's wishes, as he had the habit of refusing to publish, if not destroying, work that he felt to be inferior.

"The Days of Long Marsh," written in 1924, is the story of a lonely man who slips into the muck while walking in a tule swamp near Lomita one night. A mysterious hermit takes him to a hut and tells a sad tale: The hermit fell in love with a woman who lived in a white house on the edge of the marsh, but she had a (possibly) incestuous relationship with her brother. The hermit rigged a bridge so that the brother would fall into the quicksandlike muck, but when the woman went in after her brother, she too died. The dismal swamp perfectly reflects the narrator's mood and the hermit's melancholy story. Before the narrator meets the hermit, the setting foreshadows the tale of dread and guilt that will follow. The narrator sees his reflection in the moonlight: "And on Lomita itself there was a grey ghost, crucified, its head hanging over one shoulder." This narration seems in the romantic tradition of Poe, but Steinbeck's attention to setting is evident. In later works, such detail becomes quite powerful when it is more clearly attached to his holistic vision.

Other early unpublished stories, set in cities, make curious use of place. The cities are described in ways that make them seem more like wilderness, indicating Steinbeck's preference for natural imagery. Much of "The Nymph and Isobel," the tale of a very unfortunate relationship between a young woman and a primitive man, is set in a small Los Angeles park, described as a "little plot" which is "entirely cowed by the black cliffs of buildings all around it. . . . The small, over-tended trees had fallen into a broken hearted droop; the lawns were as worn and ugly as old green pullman seats." The setting fits the plight of the young woman, who is "cowed" by the big man so that, like the trees, she is "broken hearted." "East Third Street," possibly written in 1925-26 during or after Steinbeck's disastrous first trip to New York, is set in that city. It is an odd tale of survival of the fittest, in which a "little animal" of a man, Vinch (for finch?) surprisingly defeats Dominick, a Neanderthal bully. Like "Nymph," the urban setting is described in natural images: "A little wind boasted fitfully among the black branches of the sidewalk trees. The clouds were moon-rimmed." The wind matches Vinch's state of mind as he tries to screw up the courage, boasting to himself fitfully, to go into the bar that Dominick frequents.

"The Green Lady," Steinbeck's early 1928 draft of *To a God Unknown*, also shows the young writer's desire to connect characters to their settings. Joe Wayne, the patriarch of the story, expects his wife Beth to produce offspring to populate his ranch in Jolon. She understands contraception, however, and secretly uses it for five years while she and Joe get the ranch in order. Her apparent barrenness worries Wayne, for he is one of "those of the race which for centuries [have] fondled the breasts of the earth and cohabitated with the soil." If Beth had not produced children eventually, "he would have hated her, for his goddess was the fat pregnant earth." Later, Beth's son Andy believes his mother has been killed by the land: She slips near some ferns and dies from the fall, and he is convinced she "had been killed by the five-fingered ferns. They had been waiting to kill her." Strangely, Andy finds comfort in this knowledge about the ferns because "it was proven beyond doubt that they had a soul." Clearly, father and son in "The Green Lady" consciously connect to the land, drawing meaning and life from it. Still, in this early version, the relationship seems a transcendent one.

Connections to the land play a major role in the version of "The Green Lady" which was finally published in 1933 as *To a God Unknown*. For example, when Joseph Wayne rides out to find his friend Juanito, who murdered Wayne's brother Benjy, the shifting moods of the night are as contradictory as Wayne's feelings. The hills and the clumps of trees are "as soft and friendly as an embrace," but "the black arrow-headed pines" cut menacingly into the sky (103). The atmosphere of the land is both embracing and repelling—a parallel of Wayne's desire to find a friend in trouble and his apprehension of finding a murderer at the same time. As the "womanly" mountains reek of the curiously "pleasant bouquet" of a "skunk's anger," which from a distance smells like azaleas and adds to the mixed atmosphere of the night, Wayne nearly forgets what he is doing and begins to think of his wife, Elizabeth (103). In the grove, alone since his horse will not follow, Wayne senses evil, "for there was a breath of fear in the slumbering grove" (104). He comes to the clearing with its black rock and crawling glowworm and finds Juanito. Wayne's feelings and Juanito's situation commingle with the atmosphere of the night. Because of his mood and the impressions that the wind, trees, and hills make upon his emotions, we cannot draw a line between nature and Wayne's perception of it—Wayne and nature move in tandem.

An unpublished potboiler, "Murder at Full Moon," makes connections to place that are less romantic. In a letter, Steinbeck notes that the manuscript was written in nine days, typed up in two weeks, and is in "a slightly burlesque tone" (Benson, *True Adventures* 206). Probably written in November or December of 1930, it is possibly the first of Steinbeck's manuscripts to have been influenced by Ed Ricketts, although the novelist had only known him for a month or two. Steinbeck takes the Poesque style of the piece far less seriously than he does in "The Days of Long Marsh." In fact, Steinbeck wrote "Murder at Full Moon" under the pseudonym of Peter Pym, reminiscent of Poe's *The Narrative of Arthur Gordon Pym,* no doubt part of the burlesque of Poe's romantic style.

Despite its hasty construction and mocking tone, "Murder At Full Moon" does have a serious level, concerning itself largely with the theme of human emotion and nature. Mac, the murderer, kills dogs and people in the vicinity of a gun lodge out in the marshes near Cone City (a setting similar to that of "Long Marsh" and one Steinbeck would use later in "Johnny Bear"). Mac has a split personality, the evil half being triggered by the full moon. When Mac is caught, a psychiatrist on the scene, "Hair Doktor Schmelling," diagnoses the strange effects of the moon upon the murderer's personality. Steinbeck may have heard of the effect of moonpull on people from Ricketts. In a 1934 paper concerning tides and the animals of the littoral (reprinted in *The Outer Shores*), Ricketts discusses the work of Darwin's son, George, concerning tides and moonpull, noting the effects on marine animals and postulating that even humans are affected by "the lunar rhythm" (*Outer Shores* 2: 64). Steinbeck uses some of this essay in *The Log* (32-33).

The connection between environment and the human mind is explicit in the case of both Mac and the narrator of the story, "Egg"

Waters. Like the first-person narrators of "Long Marsh" and "Johnny Bear," Waters's moods are deeply affected by setting. He believes the murders at Gomez Marsh were brought about by the swampy atmosphere itself, for he recognizes that it makes him feel fearful and darkens his thoughts: "I am trying to put before the reader the brooding spirit of this huge area of swampland." When a trap is set for Mac, the narrator waits in the dark lodge. As his mind drifts "into melancholy channels," Waters's thoughts parallel the environment, with the wind, fog, and screeching birds outside.

In a similar vein, Luis Caré, a steward at the lodge who will die, says, "gradually the nightmare quality of the place began to work in my mind." Caré also serves to introduce a biological red herring into the mystery, for he believes the gruesome murders have been committed by some intelligent creature that has evolved in the swamp: "Suppose in the muck an evolution comparable to ours had taken place." Although there is no such creature, the story strongly suggests that when the evil in Mac's mind attaches to the dismal environment and the full moon exerts its force, a murderous beast *is* created.

The connection of fear with place, suggested in *God* and "Murder," echoes the feelings of Steinbeck and his companions in Angeles Bay in the Sea of Cortez, where they suspected the local Mexicans and Americans of criminal activity: "We have little doubt that we were entirely wrong about this, but the place breathed suspicion, and no other place had been like that" (*Log* 222). Later, Steinbeck reiterates this sense of mystery and doubts it but observes "every one of us caught a sinister feeling from the place" (223). In *The Log*, Steinbeck notes the alliance of human mood and atmosphere: "The hazy Gulf, with its changes of light and shape, was rather like us, trying to apply our thoughts, but finding them always pushed and swayed by our bodies and our needs and our satieties" (136). Such affinities are not simply metaphors; they are demonstrations of our closeness to all things, even the meteorological. A sinister mood, a sexual drive, a feeling of hatred might originate either in us or in a condition of the environment.

The same melding of human emotion and atmosphere is apparent in the snowstorm in *The Moon Is Down* (1942). The snow, in which the occupying military force will become bogged down, changes the mood of the people in the occupied town. Mayor Orden, leader of the occupied town, is pleased by the coming storm because he seems to know what will happen. The dark, gloomy weather matches a darker human mood: "And over the town there hung a blackness that was deeper than the cloud, and over the town there hung a sullenness and a dry, growing hatred" (57). During the first night of snow, Lieutenant Prackle is shot in the shoulder, revolution commences, and Orden remarks on the "sweet, cool smell of the snow" (63). The occupying patrols, heavily bundled up, march through the snow which intensifies their feelings of isolation and longing to be home. Annie, a cook among the occupied people, considers the weather: "The soldiers brought winter early. My father always said a war brought bad weather, or bad weather brought a war. I don't remember which" (80). Annie's comment underscores an interesting point: does the foul weather reflect the times, or are the people stirred to hatred by the hostility of the weather? In any case, the mood of the people and the weather are intertwined in this story, as Steinbeck suggests interrelations in the environment, the whole.

Other notable links between natural atmosphere and human mood/action may be found in *The Wayward Bus* and *To a God Unknown*, where environmental fertility parallels human sexuality; in *The Grapes of Wrath* and "Flight," where a beaten land parallels an oppressed people; in *The Pastures of Heaven*, where a bountiful land invites self-delusion; and in "The Chrysanthemums," where Elisa Allen's isolation is perfectly matched by her lonely little farm. Steinbeck intensifies such connections in his fiction through personification, anthropomorphism, theriomorphosis, and parallels of human and animal actions.

Steinbeck found these devices useful, although he knew such analogies and parallels are a "pitfall to be avoided—the industry of the bee, the economics of the ant, the villainy of the snake, all in human terms have given us profound misconceptions of the animals" (*Log* 96). "But," he adds, "parallels are amusing if they are not taken too seriously as regards the animal in question, and are downright valuable as regards humans." He knew to use these devices carefully, although they are pervasive in his work.

Personification and Anthropomorphism

Steinbeck uses personification and anthropomorphism in nearly all of his novels and short stories. These techniques subtly indicate the novelist's placement of humans in the whole. For example, in *To a God Unknown*, when Joseph Wayne first comes to Nuestra Senora, Steinbeck foreshadows Wayne's self-destructive attachment to the land in a description of giant madrone trees with "muscular limbs"; they are "pitiless and terrible" and have "cried with pain when burned" (9). Wayne believes that the great tree on his ranch, with "its old, wrinkled limbs," contains his father's spirit (25). Later, the land dies from drought, drying up the "pale hills" and "naked stones," while the black pine grove "brooded darkly" (179). These and other incidences of personification underscore Wayne's identity with the land. Indeed, Steinbeck's landscapes often take on animate forms: in *Cup of Gold* the yellow water of a river is like "a frightened, leprous woman" and caresses the hulls of boats (124); in *Tortilla Flat* the dunes near Monterey "crouched along the beach like tired hounds" (96); and in "Flight" Pepé flees in a land of "starving little black bushes" (55).

Steinbeck uses anthropomorphism to a much greater extent than personification. In *Cup of Gold,* as the Free Brotherhood—a barbarous gang of pirates—floats down the River Chagres, tigers watch the men "with a sad curiosity" while monkeys become excited "pretending to hate disturbance" and howl "their indignation" as their "right of protest" (124). In *To a God Unknown,* meadowlarks wear "yellow vests and light grey coats," hawks soar through the air "with doubled fists" (as they do in *East of Eden* also [407]), and coyotes "laugh" at the plight of a dead cow and dying calf (21, 187, 259). In *The Red Pony* chapter titled "The Promise," two horses, Nellie and Sundog, share romance in the pasture; Nellie becomes "coquettishly feminine" and afterward "her lips were curled in a perpetual fatuous smile" (*The Long Valley* 263, 265). Soldiers trudging along a lonely road in *The Moon Is Down* consider shooting a howling dog, described as "a practiced singer" who "raised his nose to his god and gave a long and fulsome account of the state of the world as it applied to him"; in other words, complaining just like his human counterparts walking along the dark road (98).

Anthropomorphism plays a part in the comedic effect of *Cannery Row*. In chapter 15, at the captain's farmhouse, the pointer bitch—suffering because her master won't take a hand in weaning her pups—looks up to Mack "saying, 'You see how it is? I try to tell him but he doesn't understand'" (199). Mack understands her and tells the captain he must wean the pups. Doc also appears to have a camaraderie with canines, for he waves at one on the street and it "smiles" back at him (213). The great frog hunt by the boys is related from the amphibian's point of view. The frogs, horrified by Mack's unfair strategy, "have every right to expect" no new game plan: "How could they have foreseen the horror that followed?" (203). "Frantic" and "frustrated," the frogs "clutched at each other" in fear and, once caught, they are "disillusioned" (204). Chapter 31 is told from the point of view of an unfortunate gopher, who finds the perfect place to construct with his "digging hands" a fine burrow complete with "emergency exits and his waterproof deluge room" (301). However, just like so many of his characters who make plans and have them dashed (such as Mack and the boys, whose plan for a party for Doc ends in disaster), the gopher must abandon his perfect habitation because he cannot find a female to share it with. *Cannery Row* is not the only work in which Steinbeck endows animals with human intelligence (for other examples, see *To a God Unknown* [8], *The Pastures of Heaven* [115], and *East of Eden* [334]).

Theriomorphosis

While personification and anthropomorphism are important elements in Steinbeck's writing, theriomorphosis is one of the most original features of his narrative style. The novelist frequently renders characters in the shape of a beast, including the apelike Allen Hueneker of *The Pastures of Heaven;* the Neanderthals, Pirate and Big Joe Portagee, of *Tortilla Flat;* the bearlike Lennie in *Of Mice and Men;* the animalistic Jelka of "The Murder;" the studhorse Victor of *Burning Bright;* and the fly/wolf Pimples of *The Wayward Bus.* This list is by no means complete. I shall consider four of Steinbeck's more notable human/beasts, Steinbeck's reminder that we still bear the stamp of our lowly origin.

Tularecito of *The Pastures of Heaven* seems both monstrous and primordial. Like early *Homo sapiens,* Tularecito's origin is "cast in obscurity." He is a creature found in the sagebrush who, as a boy, possesses "ancient and dry" eyes with "something troglodytic about his face" (44, 46). The appearance of this evolutionary throwback causes uneasiness in the people who behold him; "his discovery is a myth which the folks of the Pastures of Heaven refuse to believe" and his "troglodytic" face causes them to be "uncomfortable in his presence" (44, 46). Like an ancient man, Tularecito is closer to the earth than his modern counterparts. In his hopeless night search for gnomes, he knows the movements of deer, wildcats, and rabbits without seeing them. His closeness to the animals extends to his superior strength and his appearance; his flat face, lack of neck, and long legs cause Franklin Gomez to dub the boy "Little Frog." Gomez also refers to him as "Coyote" because "there is in this boy's face that ancient wisdom one finds in the face of a coyote" (45). As did his ancestors, Tularecito has a talent for stone carvings and animal drawings, but his artwork leads to a problem: when someone inadvertently damages one of his works, Tularecito flies into an animalistic rage.

There are echoes of Tularecito's primitive nature in the most civilized authority figures. Tularecito's rage is dangerous, animal, and immediate; Miss Martin (his first elementary school teacher) demonstrates a savage though "civilized" rage of her own. For wrecking the schoolhouse, she orders Gomez to whip Tularecito in her presence. Watching the heavy quirt fall on the boy's back, "Miss Martin's hand made involuntary motions of beating"—something vicious and untamed in her emerges in the moment of violence. (Writing out the story in a ledger book, Steinbeck made a note above this scene, probably in reference to Martin: "wants to beat him too.") Tularecito's second teacher, the enlightened Miss Morgan, also possesses primitive desires which are coaxed out by her unusual pupil. She reveres the boy's drawings and, inspired by his example, writes upon a chalk cliff she finds on her way home from school. Having pricked her finger on a thorn, she makes a mark upon the cliff with her own blood, reminiscent of the most primitive cave painters before her. Later, Tularecito's belief in old fairy tales inspires her to send him in search of gnomes, for on the blank face of the frog boy she finds "a cliff on which to carve" (55). In this story, Steinbeck's theriomorphosis of Tularecito reflects upon the others, like Miss Martin, who would set themselves up to judge this creature. Despite Tularecito's ugliness, it is this judgment upon him that makes the Little Frog such a sympathetic character. With the possible exception of Gomez, no one understands (or wants to understand) that this boy is an evolutionary link between past and present, between animals and humans.

Unsettling reflections of primitive humanity also occur in "Johnny Bear" of *The Long Valley.* Johnny Bear brings the ideal of civilized society crashing down to its vulgar reality—much to the despair of the townspeople of Loma.

Like Cone City of "Murder at Full Moon," Loma rests on a low hill rising out of the flatland of the Salinas Valley. The community reminds one of the Puritan's image of the city on the hill, for Loma's highest beacon is a church: "its spire is visible for miles" (144). But the town is often enshrouded in the "heavy pestilential mist that sneaked out of the swamp every night" (143). The narrator is there to help drain a nearby wetland to alleviate the problem. The fog creeps about the town like some animal, as critic John Timmerman points out (*Dramatic Landscape* 239). Loma, the beacon, borders a misty, primeval world similar to the one portrayed in Steinbeck's earlier unpublished works.

Johnny Bear, a frequent visitor at the Buffalo Bar, is a creature who seems to have come from the swamp. He looks like "a great, stupid, smiling bear" (146). "His black matted head bobbed forward and his long arms hung out as though he should have been on all fours and was only standing upright as a trick," Steinbeck writes. "He didn't move like a man, but like some prowling night animal" (146). This primitive beast, having no money, gets whiskey by perfectly mimicking conversations he overhears. Johnny, "the monster," is a source of fascination and apprehension for his hearers as his mimicry grotesquely parodies human beings. The narrator studies Johnny with horror: "I saw a big fly land on his head, and then I saw the whole scalp shiver the way the skin of a horse shivers under flies . . . I shuddered too, all over" (149). Like Tularecito, Johnny awakens strange reactions within people: when Johnny shudders, so does the narrator, and after watching Johnny's mimicry, the narrator mimics the bartender, Fat Carl.

While Johnny is a symbol of the beast in humanity, Loma also has its pillars of virtue. Emalin and Amy Hawkins represent the ideals of the city on the hill, as Alex, a townsman, tells the narrator: "[They] are our aristocrats, maiden ladies, kind people. . . . [T]hey aren't like other people. . . . [T]hey're symbols" (153). Johnny's imitations of the sisters have agitated Alex. It is too unsettling to have a beast mouth the conversations of the most civilized and virtuous women in town. As the narrator observes, "A place like Loma with its fogs, with its great swamp like a hideous sin, needed, really needed, the Hawkins women" (156). Alex says, "[T]hey believe in things we hope are true" (161). But Johnny's mimicry dooms the town's "symbols."

From Johnny, the people in the bar learn that Amy, much to Emalin's disgust, has had a lover and is pregnant. Amy tries to kill herself, succeeding the second time, and Johnny plays back the doctor's conversations with Emalin. Just as Johnny is going to reveal that Amy's lover is a Chinese laborer, Alex cannot stand the women's humiliation any longer, and he strikes the beast. Like Tularecito, Johnny has an animal's strength; he nearly breaks Alex's neck.

"Johnny Bear" is one of many Steinbeck tales that demonstrate the impossibility of separating ourselves from our primitive nature. A beast brutally reminds the audience at the Buffalo Bar that there are no humans above reproach and that the Hawkins sisters are symbols of Puritanical virtues that defy the biological reality of our lowly origin. Hearing Johnny's revelations, the men at the bar look "bewildered, for a system had fallen" (164). The creature also reveals that Amy committed suicide because she could not live life as a symbol; the denial of natural desires demanded of her by Emalin and the townspeople made life impossible for a healthy woman. The narrator's glimpse of Amy and Emalin along the road gives a clue to Amy's life in the shadow of her rigid, physically erect older sister: "[Amy] was very like her, but so unlike. Her edges were soft. Her eyes were warm, her mouth full. There was a swell to her breast, and yet she did look like Emalin. . . . [W]hereas Emalin's mouth was straight by nature, Amy *held* her mouth straight" (155). By sleeping with a man, Amy is fulfilling her physical needs, but the community's requirement that she uphold the celibate life of an ideal creates a deadly tension within her. The story reminds us that while there may be the ideal, an Emalin or a church spire, we must also reckon with Johnny Bear and his primordial swamp. Our connection to nature and our primitive past will be ignored at a very high price.

Simple productions like Tularecito or Johnny Bear are forms of the throwback, humans naturally close to the earth without a *conscious* understanding of their connection. Pepé of "Flight" (in *The Long Valley*), like Kino of *The Pearl,* is a human who, cast out of society, is stripped of civilization and rendered an animal before the reader's eyes. The fate of Pepé is well documented in criticism; we need only examine the theriomorphosis of this nineteen-year-old boy.

Lazy but also latently dangerous, Pepé, with his eagle nose and snakelike wrist, is jokingly described by his Mama Torres as having cow or coyote in his makeup (42). Associations of Pepé with animals continue, as he smiles "sheepishly" or when his mother calls him a "big sheep" or a "foolish chicken" (44, 45). After Pepé kills a man in a fight and flees, his animal nature is revealed by subtle narrative touches— such as the emphasis on the boy's gnawing white teeth when he eats jerky. Even more subtle is Steinbeck's blurring of pronouns in the following sentences: "[Pepé] pulled up the reins tight against the bit to keep the horse from whinnying. His face was intent and his nostrils quivered a little" (53). The "face" must refer to Pepé, but "quivering nostrils" suggests a frightened horse. Resting later, the horse is "gnawing" on dry grass while Pepé "gnawed" again at the jerky (56). When the boy encounters a wildcat, it looks at him for a long time and then walks "fearlessly" away, implying a silent recognition between two animals crossing paths. Later a mountain lion sits to watch Pepé, and moves away only when the posse arrives.

As Pepé goes along he acts less like a man, letting his horse find the direction. When it is shot, Pepé's survival is on the line, so he moves "with the instinctive care of an animal" (59). Slithering through the brush, Pepé meets a rattlesnake and a lizard on their own level. Nearing the end of his flight, Pepé's swollen tongue renders him speechless, and he has lost his gun, horse—all the implements of a man. He is a "hurt beast" who whines from his painful wound "like a dog" (64, 65). He makes his last stand as a human being, upright and "black against the morning sky." But once shot and killed, his head—the very brain and face that identify his humanity—is buried by a small rock slide. This highly detailed account of Pepé's flight, the reduction of a man to an animal, gives the story its tremendous impact. The trackers' ability to strip Pepé of his humanity is shockingly easy and convincing.

One of Steinbeck's most subtle and interesting uses of theriomorphosis is found in *East of Eden*. This novel is accurately perceived by most critics as a departure for Steinbeck, a move away from the biological perspective and toward a moralistic consideration of good and evil. Yet the character of Cathy presents a throwback both biologically and in terms of Steinbeck's work. She is a mutation, one of Darwin's "sports." Steinbeck describes her as a mental or psychic monster, created by the same process of "twisted gene" or "malformed egg" that causes obvious physical mutations (72). We may have sympathy for Tularecito or Pepé, and Johnny Bear can be forgiven as an idiot savant, but Steinbeck has reserved Cathy to demonstrate what our humanity has saved us from; she embodies the darkest side of our primitive nature. Thin, with "flaps sealed against her head" for ears, Cathy is a pretty snake (73). Samuel Hamilton notices that she does not have "human eyes"; her eyes remind him of "some memory, some picture"—perhaps satanic or perhaps primordial—the same "goat's eyes" of a bad man he had seen hanged (177, 179). As Louis Owens has observed, Cathy is "consistently described in animalistic imagery" (148).

While giving birth, Cathy cannot conceal her beastliness from Samuel. She snarls, bares her white teeth, and brutally bites Samuel's hand. "I'll have to muzzle you, I guess," Hamilton tells her. "A collie bitch did the same to me once" (193). The whole episode frightens Samuel: "This birth was too quick, too easy—like a cat having kittens" (196). Later, as a prostitute, Cathy (now Kate) at times has trouble concealing her animal nature. Once, when she indulges in too much wine, she cannot hide the beast within: "The lips of her little mouth were parted to show her small sharp teeth, and the canines were longer and more pointed than the others" (235-36). When she is sober and remembers that she has been seen in this state by Faye, her madam, Kate becomes frightened. She recalls the night's events, "moving from scene to scene like a sniffing animal" (237). When Adam Trask finds her years later as the madam of the house, Steinbeck again calls attention to her teeth, "the long canines sharp and white" (318). After she has had too much to drink, she proudly shows Adam photos of prominent men engaged in sadomasochistic sex and boasts that she has brought out the beast in them. When Adam tells her she may be "no human at all" (323), Kate replies, "Maybe you've struck it. . . . Do you think I want to be human?" (323). He notices her arthritic hand, "wrinkled as a pale monkey's paw'" (324). And when he finally triumphs, bringing out her rage, she screams "a long and shrill animal screech" (325).

With the form and selfishness of an animal coupled with a cunning and evil intelligence, Kate is an appalling creation. The only sympathetic quality Steinbeck allows her is the final realization, just before her suicide, that she is a human without humanity: "They had something she lacked, and she didn't know what it was" (553). Steinbeck draws a distinction between human and animal, and Kate must suffer the knowledge that she is on the wrong side of that line.

Parallels of Human and Animal Activity

When Steinbeck spoke of the use of "parallels" in *The Log from the Sea of Cortez*, the word may have come naturally to him because he often parallels the actions of animals with the actions of humans in his fiction, as Lisca has observed. This technique is yet another way in which Steinbeck underscores the close connections between the human and nonhuman species. Like the other techniques discussed here, these parallels appear in very early works. In "The Days of Long Marsh," the narrator hears a creature splashing in the water and some animal crying in the distance just after he has heard the hermit speak of murder. While "Egg" Waters in "Murder at Full Moon" waits inside the gun lodge for the murderer to arrive, outside the gulls scream and the drakes "complain with cracked voices," matching the narrator's own dry-mouthed fear. A more mature handling of a parallel occurs in "Flight," where Steinbeck's narration concentrates on predators and prey in the landscape while Pepé, the human quarry, flees from human hunters. The boy's trek is alive with skittering rabbits, eagles, coyotes, wildcats, and mountain lions—reminders that his flight is a drama of survival in a landscape well accustomed to pursuit and bloody capture.

In *To a God Unknown*, winter affects the animals and the rancher in the same way. After discussing the cold-weather preparations of squirrels, cows, and dogs, Steinbeck writes, "On the Wayne ranch there was preparation, too" (111). That "too" lumps the Waynes in with the other creatures and justifies Steinbeck's opening of chapter 14 with a description of the animals' readiness for winter—one sentence connects the Waynes to the whole. Steinbeck uses a similar parallel at the beginning of chapter 17, describing the spring. In a passage about the fertility of the season, and the increased number of farm animals giving birth, a simple sentence—as nonchalant as an observation of the birth of a calf—covers Alice's birthing: "Alice went home to Nuestra Senora and bore her son and brought *it* back to the ranch with her" (emphasis added, 142).

In *Sweet Thursday,* already of interest here for Doc's attempt to draw a scientific parallel between the emotions of humans and octopi, Steinbeck foreshadows the brutal battle between Doc and Joseph and Mary Rivas over Suzy. Writing about guilt, Steinbeck subtly throws out a line that sets up a later parallel: "we scream like cats in copulation" (211). Shortly after, Steinbeck describes the night of the fight over Suzy: "It was a catty night. Big toms crept about . . . seeking other toms. Lady cats preened themselves in sweet innocence" (229). That night, in the darkness, Doc finds Joseph and Mary sitting outside near Suzy's primitive boiler home; Doc attacks with a deadly fury uncharacteristic of his normally rational behavior. After Doc has recovered from his rage, Joseph and Mary cede Suzy in a most primitive (and chauvinistic) manner, tom to tom: "You ain't got any competition from me, Doc. She's all yours" (238). Even a sophisticated and scientific man like Doc finds himself acting like a tomcat on this "catty night."

Beasts at the Door

Another way in which Steinbeck connects us to other species is to lo-
cate his characters—particularly their habitations—close to other animals, one of his most pervasive techniques. The beast is always near
the door and, given the chance, it will come in. Much of the architec-
ture of Steinbeck's fiction is primitive—most often the farmhouse—
and therefore close to animals. From the ancient farmhouse described
in chapter 1 of *Cup of Gold* to Ethan Hawley's place (a sea cave) in the
last chapter of *The Winter of Our Discontent,* the animals are just beyond the stone wall or swimming into the cave at high tide. Toward the
end of *To a God Unknown,* animals move in on the ranch, "led on by
the news that the ranch was deserted" (230-231). This image of animals invading deserted human habitations is repeated in *The Grapes of
Wrath,* where the empty houses of the migrants are quickly invaded by
the animals that were always waiting just outside (126-127). What better way to illustrate the temporary nature of our separation from the
animals than to show our homes quickly occupied by other species
once we have left?

In *Burning Bright,* Steinbeck's technique becomes heavy handed
(like nearly everything else in this embarrassing work). The first act occurs in a circus tent with the sounds of lions, elephants, and horses just
outside; act two is set on a farm, with the noise of chickens, pigs,
horses, and a rooster nearby; act three lacks animal sounds, but primitive humans are represented by trophies of "war clubs," "shark-toothed
spears," and "a witch mask or two" (85). The crude settings, with beasts
outside or primitive trophies inside, underscore the brutal sexual competition that constitutes the plot.

Steinbeck makes much use of this beast-outside-the-door motif in
The Winter of Our Discontent. As Ethan Hawley sinks in stature, giving in to selfishness and greed, the change in his relationship with a
stray cat outside his grocery store emphasizes his fall. The grey cat tries
to get in the store at the beginning of the novel, but Ethan scares it off
(13). By chapter 4, when the idea of robbing the bank begins to work in
Ethan's mind, he notices the cat again and recalls that it is always there
and that he has never failed to chase it off. Ethan wonders if cats "so
resemble us that we find them curious." This time he offers it some
milk "just inside" the storeroom, but the animal leaves, apparently unable to believe Ethan's sincerity (70). By the end of part one, Ethan's

decline is well advanced; he has committed his "small rabbit slaughter" by contributing to his friend Danny's death (195). After reading Danny's will, he finds the cat brazenly trying to claw a slab of bacon, and he drives it out (200). Later, in steep moral decline after cheating his boss, Marullo, out of the grocery store, Ethan leaves milk for the cat to "invite it in to be a guest in my store" (297). Before too long he sees it lapping up the milk he has left for it in the storeroom (321). Adding to this bitter irony, his capitulation to the beast outside, Ethan learns that the body of Danny has been mutilated by some animals— "Cats, maybe," a policeman suggests (328). In his fall, Ethan Hawley finds himself in league with the animals, as selfish and brutal as they. It's one thing to desert a place and be succeeded by an animal, but quite another to stay and invite the beast in.

These techniques—setting, personification, anthropomorphism, theriomorphosis, parallels of human and animal actions, and all of the images of the beast at the door—represent the many ways in which Steinbeck emphasizes our connection to our primitive past and to our fellow species. They demonstrate the pervasiveness of his holistic perspective, so patently Darwinian. Indeed, these connections of *Homo sapiens* to the whole comprise much of Steinbeck's stylistic signature; combined, they set him apart from other writers and make up the original stamp of his prose. The narrative landscape he creates with such techniques is the perfect soil in which to plant his themes, particularly his dark picture of our species: the civilized veneer stripped off and the beast emerging in times of struggle.

COMPETITION AND SURVIVAL: STEINBECK'S VIOLENT "LITTLE SHADOW"

Violence is another Steinbeck signature; whether he writes of a simple barbecue or a workers' strike in the field, spilled blood often haunts the action. His fascination with the struggle for survival, from the creatures of the tide pool to the migrant workers of the Great Depression, is one of the most important expressions of his holism. *Homo sapiens,* when survival or possession of territory is at stake, is as subject to the law of tooth and claw as any alley cat or fiddler crab.

Critics who label Steinbeck's work as propaganda miss the point; the politics are a superficial veneer over the real drama. Steinbeck's view is timeless because at its deepest level we find the age-old struggle for existence, as Darwin saw it, and not a transitory political agenda. This quality explains why *In Dubious Battle* is no proletarian tract (or, for that matter, an anti-proletarian tract) and why *The Moon Is Down* is not typical wartime propaganda. These and other works of violence show profoundly how closely we are related to other animals, particularly as this violence is set in Steinbeck's holistic narrative environment. Occasionally offstage, often center-stage, the violence is usually there and always underscoring the fragility of those veneers—respectability and civilization—that people use to hide from what actually "is."

In *To a God Unknown,* a drought-breaking rain causes the people of Nuestra Senora to throw off all respectability and have an orgiastic wallow in the mud, to the horror of Father Angelo (18). In the last chapter, when a storm ends the terrible drought that destroyed Joseph Wayne, Angelo's religious preconceptions appear ridiculous in the face of the people's primitive nature. After thanking the Virgin for the rain, Angelo hears the wild fiesta beginning again: "'They'll be taking off their clothes,' the priest whispered, 'and they'll roll in the mud. They'll be rutting like pigs in the mud'" (263). But hearing the violent "bestial snarling" of the crowd, Angelo wisely gives up his plans to admonish them and to deny to himself once again that the naked people rolling about in the mud are animals rather than the children of God (264).

The Pastures of Heaven contains two particularly interesting stories that demonstrate the way the beast within can surface in even the most respectable people. In these stories there are no primitive Tularecitos or Johnny Bears to reflect the uncomfortable reality; rather, savagery is within the civilized people themselves.

Helen Van Deventer of chapter 5 is a Poesque figure; a well-to-do woman widowed by her husband's hunting accident, she feeds on melancholy. Because of the antics of Hilda, Helen's insane daughter, the widow moves from San Francisco to the Pastures of Heaven. She has a cabin built in Christmas Canyon, and beasts abound within and without. The living room is a memorial to her husband, replete with mounted heads, hunting guns, and a torn French battle flag—a place dedicated to slain animals, violence, and war. Helen is well aware of the creatures outside the door: the bats, quail, cows, and rabbits (she playfully converses with a little grey rabbit). The noises outside compel her to think, "It's just bursting with life" (78). When she hears a rasping sound, she assumes that animals are gnawing at the foundation of the house, the walls separating her from the beasts outside crumbling away, but the sound is Hilda, cutting through the oaken bars of the window in her room. Helen fetches a shotgun and goes after her escaped daughter. Later the girl is found dead, blown away by a shotgun blast—a grisly scene which the coroner mistakes for suicide and says of Hilda, "It was a beastly way for her to do it" (79). But, of course, it was a beastly way for Helen Van Deventer to be rid of her daughter and commence a great feast of melancholy.

Raymond Banks, of chapter 9, owns a large chicken ranch. He is a muscular man, sun baked from his work outdoors, who occasionally uses his shotgun to keep the hawks from eating up his profits. Children love him, see him as a Santa Claus, and enjoy the way he wrestles and mauls them (147). In great detail, Steinbeck describes Banks's efficient method of butchering chickens, an activity which delights the children. For vacations, he goes to San Quentin at the invitation of the warden, an old high school friend, to watch executions. Banks witnesses the killings to feed his appetite for "profound emotion" which his "meager imagination" cannot satiate (151).

At a barbecue, in which Steinbeck carefully describes the roasting and eating of chicken and beef, Bert Munroe asks to accompany Banks to an execution. Munroe, whose feelings are so sensitive that he cannot eat chicken because as a child he witnessed a botched slaughter, is startled by himself when he asks to go with Banks: "A strange thing happened to Bert. He seemed to be standing apart from his body" (155). After he hears his disembodied voice agree to go to the next execution, Munroe feels nauseated. In a classic scene, he looks at the barbecue with new eyes—it is no longer a gay social event, but a feast of slaughtered animals. He watches the grilling at the meat pits with growing agitation, a mixture of a "choked feeling of illness" and "a strange panting congestion of desire" (156). The empathetic thinking man is at odds with the old savage man; one becomes ill at the sight of dead flesh, the other aroused. Munroe goes to the table, "where his wife sat shrilling pleasantries around the gnawed carcass of a chicken" (156). The words "pleasantries" and "gnawed carcass" underscore Munroe's inner conflict.

The thinking man prevails: weeks later Munroe declines the chance to see an execution. Banks becomes angry, and Munroe explains: "If you had any imagination, I wouldn't have to tell you [why]. If you had any imagination, you'd see for yourself, and you wouldn't go up to see some poor devil get killed" (163). But Banks does have an imagination, and later he finds he cannot go to the execution either. Munroe's empathy is contagious, much to Banks's displeasure. Munroe sees the beast in himself and in others, and his perspective makes both Banks and himself uncomfortable. The barbecue scene in particular shows how empathy, Munroe's feeling for both the slaughtered chickens and the executed prisoners, can reveal the animal nature within us. To momentarily become aware of that primitive nature is confusing and painful, just as Bert Munroe is unsettled by the voice within him that wants to see another man hanged.

Similar tales that demonstrate the emergence of the animal nature of *Homo sapiens* include "The Murder" and "The Snake" (both of *The Long Valley*), *Of Mice and Men*, *The Wayward Bus*, and *Burning Bright*. Steinbeck's portrayal of the beast within, however, is most disturbing in tales of survival, when his characters flee for their lives or fight to the death. The engine of such works is always greed, the animalistic desire for a coveted object, for power, or for territory.

The Pearl demonstrates how flight and pursuit can reduce humans to an animal state. Kino, Juana, and baby Coyotito ("Little Coyote") lead a simple, natural life. In a typical Steinbeck setting, they occupy a small house with animals just outside and, like the intruding scorpion that bites Coyotito, inside as well. In the first pages they are a warm, happy family, and Kino hears the "Song of the Family" in his head, a song "saying this is safety, this is warmth, this is the *Whole*" (3). In a foreshadowing of the competition to come, a scorpion stings Coyotito, and Kino, "snarling," "teeth . . . bared and fury flar[ing] in his eyes" (6), beats and stamps this enemy. It is a struggle for life between one species and another in which Kino responds with animal rage. When he takes Coyotito to the doctor, we are introduced to his more formidable foe, the Spaniard, and in the face of this competitor Kino's response is again rage: "And as always when he came near to one of this race, Kino felt weak and afraid and angry at the same time. Rage and terror went together. He could kill the doctor more easily than he could talk to him, for all of the doctor's race spoke to all of Kino's race as though they were simple animals" (9). Thus the tension between these two races is established, ready for "the pearl of the world" to come between them and begin a deadly struggle for possession.

When Kino finds the pearl and sees Coyotito is recovering, his emotional response is to roll back his eyes and howl. In the town, "a thing like a colonial animal," the news of Kino's find spreads along the "pulsing and vibrating" nerves of the community (21). The animal in the pearl traders and even in Kino's own people—"the needs, the lusts, the hungers, of everyone"—quickly develops (21). The community moves like a great beast, the growing evil like a scorpion: the "poison sacs of the town began to manufacture venom, and the town swelled and puffed with the pressure of it" (23). The human hunting will soon commence, and in this atmosphere of growing poison, Steinbeck posits the predatory nature of fish in the sea and nighthawks in the air—a paragraph which seems included only to show the parallel between human and animal actions (33). Kino defends his home and the pearl by instinct; he is alert to the night, "the sleepy complaint of settling birds, the love agony of cats," an ancient ability that is "the gift he had from his people" (55).

In an irony reminiscent of Pepé's claims in "Flight," Kino declares himself a man after he has committed acts of violence, in this case after beating his wife for trying to get rid of the pearl ("He hissed at her like a snake, and Juana stared . . . like a sheep before the butcher") and after knifing an assailant in the night (59). He may declare himself a man, but with his inward lust for the pearl and his fear of killers and thieves, "some animal thing" moves in him, "some ancient thing out of the past" (69). Forced from the town, Kino finds himself being hunted by trackers who "scuttled over the ground like animals" (73). With hot desire for the pearl, traders and family men alike soon become savage beasts. Kino finally attacks his pursuers with animal fury and efficiency. But the death of Coyotito awakens him from his violent trance; he finally sees the pearl as a "malignant growth" and flings it into the sea—this act, rather than his fighting, releases him from the bond of the pearl and the animal it has made of him.

Such works as *The Pearl* or "Flight" pit an individual against the group; more harrowing are those stories or novels that describe competition between groups. Animals or groups will struggle most vigorously for the habitat that will best sustain them, so one of the most elemental prizes is territory. In *The Origin,* Darwin theorized that species which have developed over the widest area will be best equipped for struggle and success because they have had to overcome the most competitors. No other species is more widely distributed or more successful in the struggle for survival than *Homo sapiens* (Steinbeck makes a similar observation in *Travels With Charley* 215-16). Add to this concept Darwin's belief that "individuals of the same species come in all respects into the closest competition with each other, [so] the struggle will generally be most severe between them" (Appleman, *Origin* 115), and we can easily see that the competition among humans for survival will be the most violent.

In the same ledger book that contains the manuscript of *In Dubious Battle,* Steinbeck wrote a note to himself before he began the novel: "We are a race so filled with anger that if we do not use it all in fighting for a warm full body, we fight among ourselves. Animals fight nature for the privilege of living but man having robbed nature of its authority must fight man for the same right. Now this is necessary and all the systems fail before it." He seems to realize the darkness of this perspective, for he adds, "I have a little shadow that goes in and out with me."

If we think of *In Dubious Battle* and *The Moon Is Down* as dramatizations of a species' struggle for territory, politics fade in relevance— politics are the veneer, only an excuse for more elemental desires (land, raw materials, food). Steinbeck's own theory of the phalanx underscores the biological nature of the struggles in the two novels. Just as there is an animal, a creature of the wild past, buried within each individual, so groups of humans can become super animals.

Steinbeck's phalanx theory has been documented by many critics such as Astro, Benson, Lisca, Lester Marks (*Thematic Design in the Novels of John Steinbeck*), and most recently John H. Timmerman (*John Steinbeck's Fiction: The Aesthetics of the Road Taken*) but should be reviewed briefly here. In the two-page essay, "Argument of Phalanx," which he gave to friend Richard Albee sometime around 1935, Steinbeck describes the phalanx as a momentary grouping of individuals into "the greater beast." This "beast" has "desires, hungers and strivings" which differ from those of the individual humans that compose it. The phalanx "has memory which stretches back to the beginning of itself and to the beginnings of life"; this beast is a primitive thing, and its "emotion" may lead either to war or to art. Steinbeck is more interested in the warlike emotion. The darker side of *Homo sapiens* may form a bloodthirsty beast, or that "human" thing in us may form a good phalanx, one borne of sympathy, cooperation, "humanity." Explicit in Steinbeck's theory (and implicit in Darwin's) is that we are the most ferocious competitors among any of the species, just as the novelist's holistic perspective suggests: "All life forms from protozoa to antelopes and lions, from crabs to lemmings form and are a part of phalanxes, but the phalanx of which the units are men, are more complex, more variable and powerful than any other." In the novels *In Dubious Battle* and *The Moon Is Down*, human phalanxes grind against each other in brutal struggles for territory.

In Dubious Battle is a perfect illustration of Steinbeck's phalanx theory, that idea of which Steinbeck told friend George Albee in a 1933 letter, "There is a life time of work in it and strangely enough it is directly in line with the whole process of my other work." As Steinbeck well knew, the novel is not a simple strike novel or proletarian tract—the political implications were not that important to him. "I'm not interested in strike as means of raising men's wages," Steinbeck writes of his novel to Albee in 1935, "and I'm not interested in ranting about justice and oppression, mere outcroppings which indicate the condition" (*Steinbeck: A Life* 98). What interests him is the human self-hatred: "But man hates something in himself. He has been able to defeat every natural obstacle but himself he cannot win over unless he kills every individual. . . . The book is brutal" (98). Indeed, the novel dramatizes a species' competition with itself, the battling phalanxes of a superior fighting animal meet over the same lush territory to destroy each other. The struggle for survival is the substructure of what appears to be an ideological battle between capitalism and communism, the "mere outcroppings" of something deeper and ancient. Steinbeck's idea has interesting implications for his view of *Homo sapiens;* it puts an eerie light on such endless wars as we find in Bosnia, Ireland, and the Middle East.

From the neon restaurant sign "exploding its hard red light" in the second sentence of the novel to the abrupt cutting off of the final sentence during the speech over Jim Nolan's mutilated body, *In Dubious Battle* is thoroughly violent. As Betty Perez has shown, several critics have seen the novel as unnecessarily violent (48). It is a brutal portrayal of human self-hate, the fearsome physical struggle of man against man. In a world of beaten and crushed people, such as Jim's father or the hapless Joy, even a simple greeting resonates with violence, a sizing up of potential opponents. This uneasiness is apparent when Jim meets Mac for the first time: "Too bad we're not dogs, we could get that all over with," Mac tells Jim as they inspect each other, "We'd either be friends or fighting by now" (14). Mac decides to apprentice Jim as a strike organizer, "Kind of like teaching hunting dogs by running them with the old boys, see" (26).

Mac's dog/man similes make sense, for the characters in the novel act on an elemental level. For a strike novel, there is very little emphasis on ideology, just as Steinbeck planned: "I guess it is a brutal book, more brutal because there is no author's moral point of view" (*Steinbeck: A Life* 105). In a letter to his agent, Mavis McIntosh, Steinbeck responds to some criticism from his publisher, Pat Covici: "Answering the complaint that the ideology is incorrect, this is the silliest of criticism. . . . [I]deologies change to fit a situation. . . . In any statement by one of the protagonists I have simply used statements I have heard used" (*Steinbeck: A Life* 107). With such a view of ideologies, coupled with the desire to display *Homo sapiens* in competition with itself, no wonder the proletarian comments have small importance in the book. They are not speeches Steinbeck worked over, but fragmented recordings of things he heard tossed off. This viewpoint is reinforced by Doc Burton (one of Steinbeck's characterizations of Ed Ricketts). Doc observes that "when group-man wants to move he makes a standard. 'God wills that we recapture the Holy Land': or he says, 'We fight to make the world safe for democracy': or he says, 'We will wipe out social injustice with communism'" (131). Even Mac does not believe all the rhetoric he feeds the men, telling Jim that they have no chance of winning the strike for which the strikers will kill and die, even adding, "[I]f we won, Jim, if we put it over, our own side would kill us. I wonder why we do it" (141). Except for some vague notions that the strike may lead to better things in the future, not even Mac can clearly explain its purpose. So much for the importance of ideological movements—Steinbeck wants to cut through politics and see what truly causes people to fight and kill each other.

Rather than ideals, Steinbeck emphasizes elemental activities such as fighting, sleeping, and particularly eating. Attention to food, procuring it and eating it, occupies far more pages of the novel than ideological speeches. Why? Perhaps to remind us that the lush Torgas Valley's essential value lies in its ability to produce food. The conflict in the novel centers on food production, between those who pick the crops and those who own the land. What could be more elemental in the struggle for survival than a conflict over food and territory? The phalanx beasts clash over who will reap the most benefit from the fruits of this productive area. Another reason for the emphasis on food is to demonstrate that for all of humanity's lofty goals, such as a proletarian movement, basic animal needs will dictate success or failure.

Frequently, the strike nearly fails not for ideological reasons but simply because the strikers need food.

This elemental sensibility is enhanced by the brutal nature of food procurement. The strikers slaughter pigs and cows with bloodthirsty readiness (at one point Mac suggests they drink the blood since it's "good strong food" [213]). They need meat to energize and excite themselves for the action organized by the strike leaders. The scene in which the "hysterical men and women" crowd around the barbecue pit to watch the cooking of freshly slaughtered meat and then impatiently eat the flesh half-raw has the feeling of a Stone Age feeding frenzy (216-18). As Doc Burton puts it, laughing at Mac, "You practical men always lead practical men with stomachs" (133).

If the feeding scenes are not enough to underscore the animal nature of the group, the scenes of actual conflict certainly do. London, the most important leader of the strikers (who in fact is himself guided by Mac and Jim, the brains of the movement), commands by physical domination. His "power of authority" comes from his immense shoulders, large size, and "his dark eyes . . . as fierce and red as those of a gorilla" (47). Toward the end of the novel, when London's authority is openly questioned, the brute leader nearly tears off his challenger's jaw. The sight of blood, just what Mac had hoped for to stir the lagging enthusiasm of the strikers, not only reasserts London's authority but also galvanizes the group into action: "A long, throaty animal howl went up" (287). The sheriff's barricade that earlier held back the disorganized individuals is no match for the frenzied, blood-crazed phalanx that attacks it.

Jim Nolan is amazed by what is created: "God, Mac, you ought to have seen them. It was like all of them disappeared, and it was just one big—animal, going down the road" (288). But the mob that Jim and Mac had wanted to create comes back for Mac's blood, totally out of control after destroying the barricade. As Mac says, "The *animal* don't want the barricade. I don't know what it wants. . . . [P]eople always think it's men, and it isn't men. It's a different kind of animal" (288-289). This is the same frightening beast that Steinbeck describes in the lynching scene of "The Vigilante" or the beating of Root and Dick by anticommunists in "The Raid." The animal nature of each individual momentarily keys in, forming a collective of primitive impulses that embodies the phalanx "beast." In his concept of the phalanx, Steinbeck's notion of the human as a species is at its most disturbing.

The beast does not have to be composed of such primitive, desperate types as the impoverished strikers. *In Dubious Battle* demonstrates that, as in "The Raid" and "The Vigilante," the respectable, wealthy, and middle class can lead if not join a beast phalanx when security is threatened. London's animal nature is obvious, and Bolter, a wealthy landowner who is the new president of the Fruit Growers' Association, has his own beast—one that fine dress and a well-manicured appearance cannot quite submerge during a parley with the strikers. Despite Bolter's friendly smiles and grey business suit, the narrator observes that his "teeth were white and even" and that, when insulted, "his clean, pink hands closed gently at his sides" (221). When he smiles again, his "white teeth flashed" as if to remind us of his predatory nature.

Bolter and his kind are threatened by the revolt—their crops are in jeopardy—and so, just like the strike organizers, the owners form a phalanx with which to do battle. Negotiation is impossible, and Mac almost cheerfully observes that the strike has become a life and death struggle: "The only thing left is to drive us out or to kill us off" (259). And so, like Mac, the owners use arousing words, as in the inflammatory newspaper article quoted on pages 265-66. Vigilantes who support the owners are capable of far more effective brutality than the strikers; their snipers pick off Joy at the train station, they make Doc Burton disappear, they burn Al's lunch wagon and Anderson's barn to hurt these sympathizers, and from the shadows of trees they blow Jim's face off with a shotgun blast. If the vigilantes need assistance, the sheriff will create a mechanized phalanx of armored cars and men with grenades to wipe out the striker's camp: defenseless men, women, and children. Although the strikers do not give in at the end, for they have the new blood of Jim's body to feed them, they will no doubt lose to the more organized, more efficient, and more vicious phalanxes created by the Fruit Growers' Association. As Mac's dog/man similes remind us, for all the political overtones, the dubious battle fought over the Torgas Valley is, deep down, not so different from two packs of dogs fighting over marked territory.

"Does no one in the world want to see and judge this thing coldly?" Steinbeck asks Mavis McIntosh concerning "the situation in I.D.B." (*Steinbeck: A Life* 107). His method of seeing things coldly is to handle the "situation" as objectively as possible, without moralizing and by relying primarily on dialogue. This objective method helps to derail the political message and get down to the biological implications of his phalanx theory. What *In Dubious Battle* is to the strike novel, *The Moon Is Down* is to the war novel. Written as a play in 1942 (the novella appeared the same year), Steinbeck was caught up in the anti-Nazi fever of the times, and he does some editorializing, as John Ditsky observes in his essay, "Steinbeck, Bourne, and the Human Herd: A New/Old Gloss on *The Moon is Down*." There are lapses in the objective mode, as when the narrator of *Moon* introduces Lieutenants Prackle and Tonder as "snot-noses, undergraduates, lieutenants, trained in the politics of the day, believing the great new system invented by a genius so great that they never bothered to verify its results" (25). However, for a war story written at a time when the Nazis were winning, Steinbeck's book is remarkably free of propaganda. "It has no generalities, no ideals, no speeches," Steinbeck wrote of the play version in a letter to Webster Street. "[I]t's just about the way the people of a little town would feel if it were invaded. It isn't any country and there is no dialect and it's about how the invaders feel about it too. It's one of the first sensible things to be written about these things" (*Steinbeck: A Life* 237).

By identifying neither the sides, nor the ideologies, nor the country, and by seeing the invasion from the perspectives of both the invader and the invaded, Steinbeck is still trying to get at the whole objectively, to write "sensibly" about such emotionally charged issues as war. That his objective attempt was successful may be seen in the response to the work. Donald V. Coers, whose *John Steinbeck as Propagandist:* The Moon Is Down *Goes to War* is a fascinating historical account of the play/novella's reception, discusses in detail how the book was perceived as anything from a great propaganda piece for the war to a work that was dangerously soft on the Germans. Meanwhile, as Coers shows, the resistance movement across occupied Europe found it a source of inspiration. Once again, the biological level of Steinbeck's work confounded critics who expected any novel about the war to go only as deep as the politics of the day.

However, *The Moon Is Down* is another observation of territorial struggle, this time between two groups of a different nature—one artificial, the other organic. The invaders are "herd men" (124), units of a top-down phalanx geared to seize land, a cold, effective, and violent organization—the kind of unstoppable, coordinated beast that a strike organizer like Mac or Jim Nolan would have envied. As we see in the invaders' quick sack of the town, nothing can stop their single-minded effort when in battle. But, as Steinbeck writes in his "Argument of Phalanx," when the crisis is over and the threat to survival minimized, the units become individuals again. Hence, Steinbeck needs to show the invading pseudo-Nazis as human beings, a view which got him into much trouble (especially evident in James Thurber's angry denunciations of the novel).

The complacent natives, meanwhile, who seem so foolish and helpless during the invasion, begin to congeal into a group focused by hatred of the invaders. In the struggle for victory, even the gentlest, most civilized people may become killers. Colonel Lanser, the head of the invading force, tells of another invaded land where the hatred turned to violence: a little old woman in Brussels, with "delicate old hands" and a "quivering, sweet voice" when she sang her enemies' national song, killed twelve men with a long hat pin (a tale, incidentally, that enraged people at the Belgium Information Center of New York [Coers 17-18]). "And when we finally retreated," Lanser ends the story, "the people cut off stragglers and they burned some and they gouged the eyes from some, and some they even crucified" (41).

This same cycle of death begins in *The Moon Is Down* with the assassination of Captain Bentick, ironically the most kindly invader and a characterization that demonstrates Steinbeck's attempt to balance his work and avoid Hunnish stereotypes. Afterward, Mayor Orden explains to Colonel Lanser that the civil laws are over: "Don't you know you will have to kill all of us or we in time will kill all of you? You destroyed the law when you came in, and a new law took its place" (54). The law of survival has descended on the town. Alex, the miner who killed Bentick, goes to the firing squad, but not before Orden explains that his death will galvanize the people: "You will make the people one" (61). The pattern is similar to the one that unites the strikers, but this phalanx forms organically, in response to attack; it is not the beast of two expert manipulators. The "herd men," an artificial creation, have no staying power while the free men will eventually, as Orden says, "win wars" (124). In a pattern of survival, the group formed from natural cooperation is the more fit, despite the convulsive power of a manipulated mob. *The Moon Is Down* is less a novel about war and more an account of two kinds of phalanxes and which one is most likely to survive.

✿ ✿ ✿

So much of Steinbeck's fiction reminds us that we are animals after all. Seeing *Homo sapiens* as a species is Steinbeck's signature; no author ever explored the animal nature of humanity so thoroughly, or so scientifically, as he. The importance of this viewpoint in his work is immense—it affects his style, his characterization, and his treatment of events. This perspective, which owes so much to Darwin, is worked out darkly by the novelist. The human as animal can be disturbing; the human group as beast can be terrifying. In a world of limited resources peopled by animals of unlimited reproductive potential and constantly threatened by competition, we are in a most formidable struggle with ourselves. This is a bitter and very unflattering view, and it is no wonder that creationists still attack Darwin and that humanists deny the most essential message of Steinbeck's work.

4

Misogyny or Sexual Selection?

"Anyone who does not wish to be concerned with sex need not acquire or read this book; and those who approach the subject in a non-scientific spirit may be warned from the outset that they will find nothing suggestive or alluring in the following chapters."
—Bronislaw Malinowski, The Sexual Life of Savages, *1929*

Romance does not work for Suzy and Doc in *Sweet Thursday*, even though they and everyone else on Cannery Row are fervidly working on the relationship with all the traditional tools, from astrology to stage direction. The date at Sonny Boy's restaurant, orchestrated by Fauna, the madam of a brothel, and hoped to be reminiscent of a prom night dinner, goes well enough except that fish is served, and Suzy, allergic, is terrified of her meal. The farcically romantic Snow White extravaganza at a ribald party, also carefully arranged by Fauna, ends in complete disaster. Romance flowers only after Hazel, another of Steinbeck's primitives, has broken Doc's arm. Why can't traditional romance work even in such a sweet book as *Sweet Thursday*?

If Steinbeck's darker vision of the animal nature of *Homo sapiens* cuts through ideologies that cover our killing of each other, we should expect that his biological view of male and female relations is anything but romantic. Although *Sweet Thursday* presents love in a soft glow, Suzy's and Doc's relationship does not escape violence (recall the bloody battle for Suzy between Doc and Joseph and Mary). Never failing to probe what human as animal means, both Steinbeck and Darwin carry over their view of *Homo sapiens* into the exalted sphere of love. Because of their biological perspective, their consideration of sex is cold and brutal. Darwin's theories on human sexual relations and

Steinbeck's presentation of these theories invite criticism. Steinbeck's place in the literary canon has always been precarious, and his portrayal of male and female relations has not helped, recently earning him the label of misogynist writer. However, as in all things with Steinbeck, before we label him, we must consider the scientific implications of his work. Indeed, although his species view leads him to some unflattering descriptions of human sexuality, it also allows him to suggest ideas in a novel like *The Wayward Bus* that actually forecast Naomi Wolf's stunning work, *The Beauty Myth*.

"In the past, relatively little has been written about the female characters in Steinbeck's fiction, perhaps because these characters, in general, have been dismissed by critics as minor characters or, at best, stereotypes," Sandra Beatty writes in a 1979 collection of essays devoted to Steinbeck's treatment of female characters (1). Critics such as Beatty and Mimi Gladstein have noted the importance of his female characters, and Beatty cites Steinbeck's "many women such as Alicia in *The Pastures of Heaven*, Elizabeth and Rama in *To A God Unknown*, Ma Joad in *The Grapes of Wrath*, Juana in *The Pearl*, and Liza Hamilton in *East of Eden*" (1). To these I would add Elisa Allen of "The Chrysanthemums" (the focus of much critical attention), Mary Talbot of *Cannery Row*, Abra of *East of Eden*, and Suzy. As Peter Lisca, Beatty, and Gladstein all note, Steinbeck's women tend to be either prostitutes or mothers, and although many of these women possess wisdom, strength, and courage, to put women solely into these two categories is obviously, despite the complimentary characterizations, an extremely limited perception.

Perhaps most damning is the lack of any really independent women in Steinbeck's fiction (with the exception of the inhuman Cathy/Kate of *East of Eden*). Beatty makes a very astute observation: "[R]ather than viewing the woman as an autonomous individual, Steinbeck prefers to concern himself with women in their relationships, be they personal or professional, to men" (1). Steinbeck was not entirely blind to the meaning of his limited place for women, and the male-imposed isolation so evident in the plight of Elisa, Curly's wife in *Of Mice and Men*, or Camille Oaks in *The Wayward Bus* is sympathetically handled by the narrator. However, the limitation exists, and it contributes heavily to the perception of Steinbeck as a writer with misogynistic tendencies.

In "Steinbeck's Happy Hookers," Robert E. Morsberger writes that there "is a degree of misogyny in Steinbeck, culminating in the portrait of his one 'monster,' Cathy Ames or Kate" (36). In the diary he wrote in tandem with *East of Eden*, Steinbeck describes Cathy as both a monster and "by nature a whore" (*Journal* 39). So the novelist's most developed independent female character is both a monster and a prostitute. However, she is also another human whom Steinbeck renders in animal form—something the novelist more often reserved for his male characters. Indeed, from Samuel Hamilton's memory in *East of Eden* of the evil hanged man he once saw, we know that Cathy's type includes male human monsters as well. In his diary, Steinbeck explains why Cathy had to be a woman: "If she were simply a monster, that would not bring her in. But since she had the most powerful impact on Adam and transmitted her blood to her sons and influenced the generations—she certainly belongs in this book" (*Journal* 42). The plot of *East of Eden* made it necessary that this creature be female.

The most serious examination of Steinbeck's misogyny occurs in "*The Wayward Bus*: Steinbeck's Misogynistic Manifesto?," by Bobbi Gonzales and Mimi Gladstein. This essay, although challenging and engaging, operates from the debatable thesis that the misogyny of the novel arises from Steinbeck's personal troubles during his short marriage to Gwyndolyn Conger: "[E]ven a cursory reading of *The Wayward Bus*, written during two of those four bitterly unhappy years, will reveal its author's less than salubrious attitudes toward women, male/female relationships in general, and marriage in particular" (157).

The facts simply do not support the notion that Steinbeck wrote the novel as some pointed attack on his wife or on women in general. As Gonzales and Gladstein acknowledge, *The Wayward Bus* was dedicated to Gwyn, and their "relationship seemed satisfactory and secure for, during the process of writing the book, their second son John was born" (157). In a letter written while he was working on the book, Steinbeck reported that Gwyn liked what she had read of the novel and, along with Pat Covici, was the only one to have seen any of it (Benson, *True Adventures* 582). How could she have missed an attack upon her in a manuscript being written by her husband?

The Wayward Bus came out in 1947, the same year as *The Pearl*, in which we find Steinbeck's Juana who, as Gladstein writes in "Steinbeck's Juana: A Woman of Worth," is "after Ma Joad . . . the most positively depicted woman in Steinbeck's works" (49). Immediately after his divorce of Gwyn in 1948, during the period in which Steinbeck was most bitter, he wrote the screenplay for *Viva Zapata!* and began the play/novella, *Burning Bright.* Beatty finds the women in these productions, Josefa in *Zapata* and Mordeen in *Burning Bright,* to be individual and complex creations with some of the most positive characteristics ever portrayed in Steinbeck's women (7).

During the late summer and early fall of 1948, Steinbeck was living a self-described nightmare at the cottage in Pacific Grove, and Mildred Lyman, of McIntosh and Otis, Steinbeck's agency, noted his strange views about women (his letters during the period *do* reveal misogynistic attitudes). But he seems to have refused to let the venom get into his work. Writing to friends Joe and Charlotte Jackson, Steinbeck tells them that he wrote a "completely evil" story that was very effective—but he immediately burned it (*Steinbeck: A Life* 336-337). Perhaps this story was his misogynistic manifesto, but in any event, he knew it to be unsuitable for the public. (Similarly, when Steinbeck wrote a mean-spirited satire of depression-era California, "L'Affaire Lettuceberg," he recognized it as a creature of his anger and destroyed it, writing *The Grapes of Wrath* instead.) The letters, where he vented his fury to his friends, were never meant for publication; like the wicked story he wrote and destroyed, they represent the kind of venting typical of anyone dealt a bitter divorce and the death of a close friend (Ed Ricketts).

In fact, one of his confidants during this period was Wanda Van Brunt, a friend who was suffering through an ugly divorce of her own. The two commiserate in letters, and Steinbeck writes, "Thank you for your help and your encouragement when my flag was down. . . . It's a good thing to have friends" (12 Sept. 1948). In early 1949 he shows that he can be as critical of husbands as he has been of wives and that he is not above blaming himself either: "It seems to me that in two cases you had children instead of husbands . . . Both of us seem to have married children and maybe that's what both of us wanted" (16 Jan. 1949). Despite Steinbeck's angry attacks on his ex-wife and occasionally women in general, he hardly seems a misogynist in his correspondence with Van Brunt. While he was still very angry, he wrote to her that "the institution of marriage isn't one I am good at. I guess I will give it up for good" (28 Sept. 1948). The views expressed in his letters of this period are transitory; he married his third wife, Elaine, in December, 1950, and remained very happy with her until his death in 1968.

Although certainly tension existed between John and Gwyn during their four years together, little evidence suggests that Steinbeck chose *The Wayward Bus* to be his misogynistic manifesto. If he had chosen to allow his anger against women to spill over into his creative work, most likely he would have done so in 1948 or early 1949 (before he met Elaine) when, as his letters show, he was most depressed and angry about the divorce and about women in general. Probably Cathy is his most misogynistic creation, more so than any character in *The Wayward Bus*, and yet he put her on paper at a time when his personal life was apparently extremely pleasant.

But the Gonzales/Gladstein article does raise some important questions. What can be made of *The Wayward Bus*, in which the "women are seen by the men in their lives as serving only one purpose: they are sex objects, objectified to the extreme, there to satisfy a basic animal drive," and in which "none of the sexual couplings in the novel is contextualized in a way that could be characterized as loving" (159). If there is a streak of brutality in Steinbeck's darker view of male/female relationships, it is an extension of his biological view of *Homo sapiens*. We must look at the biological substructure of his art; in matters of sex, particularly in works like "The Murder," *Burning Bright*, or *The Wayward Bus*, his views parallel Charles Darwin's.

Steinbeck could have learned of Darwin's theory of sexual selection from Smuts's summary in *Holism and Evolution* or from the brief sketch in *The Origin of Species*, but whatever the source, Steinbeck did extend his biological view into human sexuality just as Darwin did, and they hit upon similar conclusions. The novelist's thinking might have been further refined by Briffault (Steinbeck read *The Making of Humanity* and *The Mothers: The Matriarchal Theory of Social Origins* [DeMott, *Steinbeck's Reading* 18]), who was well steeped in the theories of Charles Darwin. Looking at these Darwinian conceptions today, we find them objectionable and their dramatization in Steinbeck's fiction downright disturbing.

Darwin's theory of sexual selection extends competition to sexual matters and pertains to all species, including *Homo sapiens*. When he writes, "On the whole there can be no doubt that with almost all animals, in which the sexes are separate, there is a constantly recurrent struggle between the males for the possession of the females," he means *all* animals (*Descent*, Modern Library 572). The most successfully competitive male will dominate, therefore winning the most desirable female. In generation after generation, the strongest of the species will win the right of procreation. While the strongest male will likely fertilize a greater number of females, the weaker males will be less likely to breed. Selection is further facilitated as the best females will be attracted to the best males. Over time, characteristics such as size, strength, and adornment will be selected. Since the burden of attracting and winning a female falls upon the male, selection will have the most profound effect on him. Darwin thus explains why the male typically has the greatest size and strength and the most showy adornment.

Regarding humans, Darwin writes, "There can be little doubt that the greater size and strength of man, in comparison with woman, together with his broader shoulders, more developed muscles, rugged outline of body, his greater courage and pugnacity, are all due in chief part to inheritance from his half-human male ancestors" (*Descent,* Modern Library 872). In his consideration of the effects of sexual selection on humans, Darwin's writing seems never more open to debate. He feels that sexual selection has played an important role in what he describes as the differing mental powers of men and women. In other animals, such as pigs, cattle, or apes, he asserts, the dispositions of male and female are markedly different. Among humans, "Woman seems to differ from man in mental disposition, chiefly in her greater tenderness and less selfishness. . . . Woman, owing to her maternal instincts, displays these qualities towards her infants in an eminent degree; therefore it is likely that she would often extend them towards her fellow-creatures" (873). "Man is the rival of other men," Darwin writes. "He delights in competition, and this leads to ambition which passes too easily into selfishness. These latter qualities seem to be his natural and unfortunate birthright" (873). Darwin finds man's mental capacity generally superior to woman's, due to the sharpening of competition, so that the "chief distinction in the intellectual powers of the two sexes is shewn by man's attaining a higher eminence, in whatever he takes up, than can woman" (873). For Darwin man is the competitor, the striver, while woman is the vessel of compassion, the maternal life preserver.

If Steinbeck had not imported these "natural" roles of men and women directly from Darwin, the novelist would have read similar views (without the emphasis on sexual selection) in Briffault's *The Mothers.* "Had the incipient human social group been a herd or horde ruled by the selfishness of a despotic patriarchal male," Briffault writes, "[human progress] would have been subject to a heavy, if not indeed a fatal, handicap" (66). For all of Ma Joad's selflessness, wisdom, and courage in *The Grapes of Wrath,* in the end only Tom and Casy can see the big picture; she can hardly see beyond the circle of the family. Indeed, Steinbeck's first wife, Carol Henning, has said that Ma Joad was "pure Briffault"; the novelist was influenced by Briffault's belief in the importance and power of "maternal impulses" in "the animal family" (Astro, *Steinbeck and Ricketts* 133).

Steinbeck's women, typecast in certain roles and largely supporters of bolder men, are conceived by the novelist from a biological generalization rather than a conscious disdain. Certainly the novelist's Darwinian, biological conception of man and woman is frequently demonstrated in his fiction.

The view of woman as a selfless, maternal savior of men, a fine but only a *supportive* role, is very clear in *The Pearl*. After Kino beats Juana for sensibly trying to get rid of the pearl, the narrator discusses the difference between men and women from Juana's point of view. Kino not only beats her, but he also fights off an unknown male assailant. Kino's sense of himself is sharpened by this bloody competition and, reminiscent of Pepé after he kills, Kino tells Juana, "I am a man" (59). To Juana, this means her husband is "half insane and half god" so that "Kino would drive his strength against a mountain and plunge his strength against the sea" (59). Juana knows that such superhuman ambition may break him, but this is what she needs in a man. The narrator adds, "Sometimes the quality of woman, the reason, the caution, the sense of preservation, could cut through Kino's manness and save them all" (60). A man like Kino, although selfish in his greed for the pearl, has great potential in that he risks reaching for things that are beyond him. He is aggressive, a competitor; Juana, as a woman, is more selfless and maternal (she clutches Coyotito throughout the chase as Kino defends them). Her sense of preservation means that she can toss the pearl away (only after Kino agrees, however) but that she will not strive for greatness. For all of Juana's fine qualities, like Ma Joad, there is a plane of male achievement that is beyond her earthly horizon.

In a more limited range, Steinbeck's women exist on a kind of evolutionary scale between beast and human just as his men do. The males in Steinbeck's works exist in far greater numbers, and given a Darwinian assessment of their biologically designated potential, we have characters such as Johnny Bear and Tularecito on the low end and Tom Joad and Doc on the high end. The range for females is more circumscribed. Those women who most resemble animals do not fall as low as their male/beast counterparts. Characters such as Lisa of *In*

Dubious Battle or Jelka of "The Murder" should be thought of not as representatives of Steinbeck's latent misogyny, but rather as extensions of his larger portrayal of *Homo sapiens* in general.

As Gonzales and Gladstein rightly point out, Lisa "is painted in bovine imagery" with mental powers so limited that "her one desire, amidst all the turmoil and violence of the strike, is for a cow so she could make butter and cheese" (158). But her inability to see beyond a yearning for food simply puts her in the same bovine class as most of the male strikers—she is a good cow, looking after her young, while the bulls outside senselessly thrash each other or gather to battle the other herd in the next pasture. Once again we see what Steinbeck meant when he said *In Dubious Battle* is an ugly book that no one will like: it is a brutal, unflattering representation of the most base elements within us.

Steinbeck's Darwinian view of women and sex can be best studied by looking at those few works that focus on male/female relationships. "The Murder" portrays his most loathsome and controversial relationship. *Burning Bright* is an overt, terribly clumsy picture of his biological view of human sexuality. *The Wayward Bus*, however, is his most sophisticated look at male/female relationships, and might be seen as either a misogynistic manifesto or an intelligent, prophetic novel.

"The Murder"

As Robert Murray Davis has observed, critics have avoided comment on the disturbing characterization of Jelka, a woman who wants to be beaten by her husband (63). According to Benson, the story is based loosely on an actual case in which a husband killed his wife and her lover, and the character of Jelka may have been derived from the "peculiar characteristics" Steinbeck saw in a friend's wife (*True Adventures* 274). Another source may have been discussions of the violence of love among primitive cultures by Darwin and Briffault. Considering Jelka's European background, a comment by Briffault in *The Mothers* is interesting: "In many parts of Europe women are not convinced of their lover's or husband's affection unless their own bodies bear the visible marks of it in the form of impressions from their teeth" (49). But still, why did Steinbeck wish to set down such a reprehensible tale? Jelka, like so many of Steinbeck's base characters, is an ugly demonstration of human animal nature.

Jim Moore and Jelka live in an old ranch under the shadow of a rock formation, a "tremendous stone castle" like one of "those strongholds the Crusaders put up in the path of their conquests," replete with "ruined battlements" (170). As in most of Steinbeck's fiction, place means something: here we have the suggestion of an ancient European battleground, just the place for the bizarre old-world relationship that is played out in the house Moore eventually abandons. He marries Jelka Sepic, a Slavic woman. Her drunk, "bleary and bloated" father tells Moore to beat her just as the old man beat her mother (171). This is obviously a disgusting old custom, and Steinbeck's opinion of it is reflected by the grotesque character he chooses for its representative.

Jelka is primitive, speaks little, and is so animal-like that Moore pats her head and neck "under the same impulse that made him stroke a horse" (172). She is uncommunicative, and Moore takes to chatting with the noisy girls at the Three Star, where he always jokes that his wife stays "home in the barn" (173). On a night lit by a full moon, Moore leaves his strange wife, but, learning from a friend that rustlers are about, the rancher comes home early to protect his livestock—ironically, not the only "animal" he must see to that night. Moore finds his wife sleeping with her cousin, and he creeps out, overwhelmed by black thoughts of slaughter and blood. He goes back to the room and shoots the cousin in the forehead. Jelka's reaction is to whine "softly, like a cold puppy" (181). Outside, his rage spent, Moore vomits, ill with the realization of what he has done. After the deputy and coroner leave, Moore finds Jelka—still whimpering like a puppy—and strikes her savagely with a bullwhip. She seems favorably impressed, saying, "You hurt me bad" (183), and when she goes to fix him breakfast, "[h]er dark eyes dwelt warmly on him for a moment" (184). She asks, in a most hopeful way, if he will beat her again, and he replies: "No, not any more, for this" (184). The story ends with smiles in Jelka's eyes as Moore strokes her hair and neck.

Certainly Steinbeck does not want us to applaud this relationship. Rather, this story, rendered in disgusting detail right down to the mucus running from Jelka's nose during the murder, holds up the possibility of a dark, ancient savagery still alive in the sexual relationship of this twentieth-century couple. Most chilling is that Moore, a civilized man, easily succumbs to the violent love of Jelka's old-world heritage. The story is a stark reminder that the animal traditions of our primitive past may well overcome the codes of civilization, much as the animalistic Johnny Bear brings down the cultured aspirations of a town. In "The Murder," Moore learns that to keep his strange wife, he must demonstrate his ferocity and strength—in a sense, live by the tenets of Darwin's sexual selection, as a bull must to keep his cows. The irony is that the civilized Moore is brought down so easily by the animalistic Jelka, and the episode is an example of Steinbeck's frequent examination of the fragile line between human and beast. Although disturbed by this trend in Steinbeck's work, Edmund Wilson is correct when he writes, "Mr. Steinbeck does not have the effect . . . of romantically raising the animals to the stature of human beings, but rather of assimilating the human beings to animals" (43).

Burning Bright

"'We didn't invent this—it happens every day,'" Victor tells Mordeen in *Burning Bright* the night he will mount her to stud a child for her husband, Joe Saul (49). In this play/novella, no doubt the worst piece of writing Steinbeck ever published, the characters find themselves in a situation almost as primitive and loathsome as that of "The Murder." Saul, nearing fifty, has been married three years to his young wife, Mordeen. She learns that they cannot have children because of Saul's impotence, although he is not aware of his problem and is desperate for a child to carry on his bloodline. Meanwhile, Victor, a virile, strapping young man, senses the inadequacy of the older male and stalks the young female. The situation seems locked in a natural pattern of sexual selection—we might expect Victor to defeat Saul (at one point in the first act they come close to fighting) and take Mordeen away with him.

However, this animal pattern is altered because of Mordeen's capacity for savagery as well as love; where in "The Murder" Jelka pulls Moore down, Mordeen first tries to pull the men up but then falls to their level. Still, within the confines of her "natural" role, Mordeen succeeds. As Beatty has noted, "Through her unselfish devotion to her husband, she has succeeded in bringing to him the joy and contentment she promised and thus is truly fulfilled as a woman" (12). Mordeen, although admirable for her strength, is still restricted: typical of Steinbeck's women, she serves her man—in this case, by fulfilling her biological function (just as Beth did for Joe Wayne in "The Green Lady"). Mordeen allows Victor to act the stud and afterward tries to discard him.

Victor can never see beyond his selfish desires, particularly his animalistic lust for Mordeen. However, we can sympathize when he realizes he has been exploited and confronts Mordeen: "Do you think I want to be used like a stud animal for the comfort of Joe Saul? Is that fair? He gets everything, and I get put back in the corral" (69). But she has already forgotten his part and is bent on her maternal role. She puts up a protective wall between Victor and the unborn child: "She glared at him like a mother cat, and her claws were out. And then she backed to the cot, her teeth bared and her nostrils flaring" (70). Mordeen tells him that he can find his own true love, the kind of affection that is between Saul and her, but that for now she must protect her family and so he must go.

In act three, Victor returns for her and the child. He cannot see beyond his own wants: "I'm in a long narrow tunnel and I can't turn" (91). When reason cannot overcome Victor's desire, Mordeen resorts to violence—she draws a short knife from the wall in the ship's cabin and hides it, waiting for her chance to slay him. Fortunately for Mordeen, Friend Ed, who as a "Doc" character (another fictionalized version of Ed Ricketts) has been a highly rational mediator throughout the play/novella, escorts Victor outside, kills him with a "crunching blow," and tosses the body over the side. In a very lame bit of dialogue, Mordeen says to Ed, "He was not evil," and Ed responds, "I know" (93).

Joe Saul then arrives, enraged, since he has learned from Dr. Zorn that he cannot have children. Friend Ed tells him to see beyond his "smallness" and welcome this great gift of love that Mordeen has bestowed upon him (198). In the final scene, Saul learns to accept Mordeen's gift and see the big picture: "It is the race, the species that must go staggering on" (105). For all of Mordeen's nobility as a character, she still remains in the biological role of progenitor, mother, and thus servant to husband and child.

Burning Bright is badly flawed in many ways (the experimental language with which Steinbeck tried to convey his Everyman message is a disaster), but the greatest failure is that what the play/novella might have been meant to say—that the human capacity for love and compassion can overcome animal baseness—collides with the actions of the characters. For Saul, Mordeen is a selfless vessel of courageous love, but for Victor she is a manipulative mother cat. In act two, Victor, the hapless stud who out of loneliness longs for a higher love with Mordeen, plaintively tells her, "I love you. And it's not like anything I have ever known" (71). For his pains, he gets a crushed skull and a boot into the sea (he would have been knifed if Mordeen had reached him before Ed). Friend Ed's solution to Mordeen's dilemma does not require much depth of mind.

We cannot accept Steinbeck's lofty claims of love and compassion in this play/novella when those who speak of these exalted qualities are accomplices in a brutal murder. Steinbeck wants to show how a higher vision can break the narrow tooth-and-claw view, but the murder of Victor smells of a survival act of animal violence. Indeed, since Saul learns of his impotence and then quickly accepts his wife's inconstancy, the murder proves unnecessary. Nevertheless, Mordeen clearly is meant to be a positive character despite the limitations inherent in the author's biological viewpoint.

The Wayward Bus

The best place to find Steinbeck's expression of the sexual human is in *The Wayward Bus*, a novel conjuring up the author's darker view of *Homo sapiens*, to be sure, but hardly a misogynist manifesto. What *In Dubious Battle* is to conflict, *The Wayward Bus* is to sex. It is the story of people tightly wrapped in the lies of a commercial society, lies that deny or cosmetically gloss over even such an honest and basic biological act as sex. In a spring environment charged with fecundity, the bus takes a wrong turn, stalls in the mud, and the repressed animal desires of a group of characters are loosed in a primitive environment. Both the men and the women represent points on a scale: those who recognize the commercialized illusion they have been living rank highest; those who cannot fall low, little better than frustrated animals. This stark portrayal of the naked *Homo sapiens* is not romantic and certainly not beautiful.

Like *In Dubious Battle*, *The Wayward Bus* is unpleasant—as Steinbeck knew it to be. Before writing the book, the author thought he would create something like *Don Quixote* about "a cosmic bus holding sparks and back firing into the Milky Way. . . . *The Wayward Bus* will be a pleasant thing" (*Steinbeck: A Life* 284). His opinion of the novel changed after its completion, and his comment about it to friend Jack Wagner nearly two years later echoes his thoughts about his ugly "strike" novel: "I hope you will like it although 'like' is not the word to use. You nor anyone can't *like* it. But at least I think it is effective" (*Steinbeck: A Life* 296). If, as Gonzales and Gladstein assert, *The*

Wayward Bus is an ugly look at "Everywoman" (159), it is an equally unflattering look at Everyman. Steinbeck spares neither sex.

Ironically, for a novel that at first glance has a brutality that rings of misogyny, *The Wayward Bus* forecasts important feminist revelations about human sexuality that culminate in Naomi Wolf's *The Beauty Myth*. Steinbeck's frank look at sexual delusion and oppression in America is an extension of his biological perspective, of the recurring theme in his work that truth is often hidden by civilization. At times, particularly in his characterizations of Mildred Prichard, Camille Oaks, and Norma, the novelist's denunciation of the artificiality of a commercialized society jibes with Wolf's thesis.

The Beauty Myth attacks the oppressive, highly artificial twentieth-century conception of women's beauty. Forcing women to aspire to impossible physical norms created to maintain high sales and profits by a largely patriarchal industry—a conglomeration of advertisers, fashion designers, cosmetics corporations, and plastic surgeons—this myth drains women of health as they starve or mutilate themselves to achieve the artificial ideal they see in magazines and on television. Robbed of self-esteem, they must consider sex from the point of view of men and reflect what the myth tells them men desire. It forces women to compete against each other and dislike those who come closest to the norm. And it makes women's independence extremely difficult—one who spends fortunes and hours applying makeup or fixing hair or who becomes obsessed with restrictive diets and frequent surgeries simply lacks the time and health necessary to compete in the corporate or political world.

Wolf's book is actually an extension of Betty Friedan's groundbreaking 1963 critique of American society, *The Feminine Mystique*. "Feminine mystique" is Friedan's term for the quiet oppression of women directed to the one role of housewife, a particularly powerful social drive in the '50s and '60s. She notes that women in America are confined to their biological role, trapped in a "world of bedroom and kitchen, sex, babies, and home" (36). Like Wolf, Friedan faults advertisers and the media in general, which enforce the biological role with an image that "shapes women's lives today and mirrors their dreams" (34). This image makes an ideal of women who "do no work except housework and work to keep their bodies beautiful and to get and keep a man" (36). Pushing the feminine mystique, Friedan observes, is the business of advertisers because it helps increase sales in everything from cosmetics to appliances (72). She criticizes Sigmund Freud's work for contributing to the feminine mystique, particularly his ideas that anatomy is destiny (your sex determines who you are) and that women suffer penis envy (24, 114). She also criticizes the work of anthropologist Margaret Mead, whose work more and more became "a glorification of women in the feminine role—as defined by their sexual biological function" (137). (Steinbeck might have been acquainted with Mead's *Coming of Age in Samoa* and *Growing up in New Guinea,* as these became part of Ed Ricketts's library in 1937 [DeMott, *Steinbeck's Reading* 79]). Although Friedan does not discuss Darwin's influence, his theory concerning sexual selection and the limiting biological role of women was another great cultural resource for the feminine mystique.

Of course, *The Wayward Bus* is a product of its times, and Steinbeck's Darwinian viewpoint is at odds with both Friedan and Wolf. Disagreeing with the principles of sexual selection, Wolf points out that ideals of beauty "change at a pace far more rapid than that of the evolution of species" (12). The beauty myth is not a product of evolution but is born of "politics, finance, and sexual repression" and "men's institutions and institutional power" (13). Wolf further recognizes that the pains taken to be beautiful today, especially through cosmetic surgery, cannot be justified by "the fallacy that beauty is a form of Darwinism, a natural struggle for scarce resources" (236). Actually, on this latter point, Steinbeck might have agreed that mutilation has little to do with recognizing natural processes or biological reality, and his desire to tear down the commercialized veneer that obscures natural sexuality comes close to Wolf's and Friedan's attempts to explode the beauty myth.

But while Wolf's examination of human sexuality concerns such areas as finance and politics, Steinbeck's perspective is typically biological. For example, his usual linking of human mood and natural environment occurs in *The Wayward Bus*. Gonzales and Gladstein note that "[t]he book's original title, 'Whan that Aprille,' suggests that Steinbeck had in mind a Chaucerian medley of pilgrims, a representative microcosm of our society" (158). Considering Steinbeck's more obvious use of the Everyman theme in *Burning Bright* three years later, their assertion seems especially astute. However, "Whan that Aprille" also underscores the important sexual environment in which the "pilgrims" find themselves. This unused title comes from the first three words of the prologue of *The Canterbury Tales*. The first seventeen lines are full of rich sexual imagery, as many critics of Chaucer have observed. Arthur W. Hoffman's classic commentary on the natural (animal) and supernatural (spiritual) implications of the prologue quickly identifies the opening sexual imagery: "The phallicism of the opening lines presents the impregnating of a female March by a male April" (31). The natural environment of *The Wayward Bus* is similar; Steinbeck parallels the climate, an eroticized spring, with the actions of his characters, a group burning with pent-up sexuality. "The sweet smell of the lupines and of the grass set you breathing nervously," Steinbeck writes near the beginning of the novel, "set you panting almost sexually" (8). Toward the end he emphasizes this image: "a new breeze had come up, bearing the semenous smell of grass and the spice of lupine" (210). The "semenous" image again recalls Chaucer's prologue, in which April bathes March "in swich licour" (17).

A look at the individual characters in the novel and their sexual relationships is most revealing. The strongest characters know their sexual potential and control it, the weakest are slaves to sexual impulse or to a commercialized vision of what sex should be.

Gonzales and Gladstein rightly observe that Juan Chicoy is "The touchstone of masculinity in the novel" (166). Although past his prime (about 50), he remains in fine physical shape, drives the bus, and is in control of his own sexuality. Alice Chicoy is "insanely in love with him and a little afraid of him, too, because he was a man, and there aren't very many of them" (*Bus* 3). His complete domination of his wife is the best argument for a misogynist viewpoint on the part of the author. "[A] fine, steady man," Juan nevertheless is probably the lowest among those we might identify as Steinbeck's heroes (3). The novelist does not endow Juan with the same intelligence and consideration as the "Doc" characters, nor with the vision and sympathy of Tom Joad or Casy, nor the plain sense of camaraderie and fun of the paisanos. Like everyone else in the novel, Juan is frustrated and lacks the imagination to do much about his situation. He lives by the charms hanging over the dashboard of his bus and prays to his metal Virgin; without a clear idea of his place in the universe, he does not have any profound answers and feels dissatisfied with his life.

Ultimately, Juan (like his wife) is trapped in Rebel Corners, sentenced to run the bus on its little route year after year. Juan's compassion is very limited, as demonstrated by his coldness toward his wife, whom he has conditioned to enjoy his beatings, when he "hit her as he would a bug" (*Bus* 23). The Chicoys function on that animal level that Steinbeck has never portrayed as the most admirable place for his characters. Juan has his narrow perceptions; "an imp of hatred" stirs in him when he perceives Mildred as representative of the white race that suppressed his ancestry (56). His only plan is to escape Rebel Corners, and he thinks of wrecking the bus on the muddy road, perhaps just walking away from it forever. In his effort to get away, he goes as far as a barn down the road and finds—like Pat Humbert of *The Pastures of Heaven*—that his course is utterly circumscribed. The Virgin does not help Juan, and he must return to his life. Simply put, Juan Chicoy is not one of Steinbeck's grand heroes, although he does have control over his sexual impulses, understands them, and therefore places high on the scale of sexual selection, rather like a seasoned stallion.

If Juan and Alice were the only characters of *The Wayward Bus*, it might indeed be a misogynist manifesto. However, there are women in the book who equal Juan, if not surpass him, in potential and in sexual power. Juan eventually seduces Mildred Pritchard, but not without her reciprocal selection of him. Young, strong, and intelligent, Mildred has a healthy sexual appetite which she is not afraid to satisfy (she has consummated two affairs). Mildred is still largely inexperienced and has "variable" convictions, although "she had undertaken causes and usually good ones" (44). She is one of the few characters in the novel whose life has not settled into a pattern; she suffers from few illusions. Unlike her mother, Mildred has suffered no "physiological accident" and so enjoys her sexual desire; she feels a yearning for Juan, knows it to be "crazy," but gives in to the "quick and sexual picture [that] formed in her mind" (55).

Disgusted with her parent's quarreling, Mildred walks down the road in the same direction as Juan and with the same idea in her mind: "Suppose Mildred didn't come back? Suppose she walked on and caught a ride and disappeared?" (175). She follows Juan's tracks. Along the way, she recognizes her strong sexual desire, knowing that to satisfy it "she would either have to get married or make some kind of permanent arrangement" (176-77). She searches for Juan to find some temporary relief, already knowing Juan's limitations, wondering if, like her father, he might be "held in line" by "habit" (179).

Upon meeting, they lie to each other—Mildred says she has just been out walking to get exercise; Juan explains he is in the barn to get some rest. What ensues is a kind of perfunctory lovemaking, a mockery of all the trumped-up preliminaries in movies and magazines: "Aren't you going to make a pass at me?" Mildred asks; "Yes, I guess so," Juan responds (182). At first, Mildred is open about her desires, telling Juan that her willingness to stay with him in the barn has nothing to do with him: "I know what I want. I don't even like you" (181). She also tells him that she feels alone: "I can't tell anyone anything" (181). But as they come closer to having sex, she falters, asking him to force her "a little." "I'm what you'd call an intellectual girl," she tells him. "I read things . . . but I can't make the advances" (182). Juan understands, observing that the women in Mexico need to be begged or forced (an unpleasant view reminiscent of "The Murder"). Part of this ritual requires that the man must want the woman, and Mildred tells him she cares about his desiring her "whether I want to or not" (182).

These artificial preliminaries constitute yet another cultural gloss over sex in a novel that consistently derides such things. The cultural necessity of being forced is portrayed as something disgusting in "The Murder." In *Burning Bright*, Victor makes blunt advances on Mordeen, claiming that "every girl in the world needs—a little bit of forcing" (37). Mordeen doesn't react as Victor expects, and when she stubbornly becomes dead weight in his arms, he is "puzzled" by her resistance (37). She backs away from him, feeling "hatred and contempt" (37). Mordeen, one of Steinbeck's strongest female characters, knows much more about sex and love than Victor ever will.

In *The Wayward Bus*, Mildred is not as programmed in the ritual as Jelka, but she is not experienced enough to reject it as Mordeen does, either. Juan clearly knows the thing is a game, and he mocks it and Mildred as well. The necessity of force in a relationship comes from a woman's need to be wanted and to be freed of guilt for having unsanctioned sex; as Juan says, after being begged or forced, women "feel good about it" (182). Obviously, this is a gross perversion of natural sex. Indeed, Darwin believes that sexual selection is a reciprocal process: while males of a species fight for the right of procreation, "the females have the opportunity of selecting one out of several males, on the supposition that their mental capacity suffices for the exertion of a choice" (*Descent*, Modern Library 571). In keeping with the damnation of artificiality that is at the heart of *The Wayward Bus*, the ritual of force is not applauded; it is yet another lie which bars someone like Mildred from easily acting on her natural desire.

Steinbeck's exposure of this ritual forecasts Friedan's feminine mystique and Wolf's theory of the beauty myth. Despite Mildred's recognition of desire, her boldness with Juan, and her intelligence, she stills plays into the myth that dictates that a woman should have no desire of her own. With no one to talk to concerning her interest in sex, and no doubt with the specter of her Puritanical mother hovering in the background, Mildred falls back on convention: she needs to be forced a little, needs to be desired by a man, even one she doesn't really like. As Wolf writes, the beauty myth shifts all power of desire to men so that women are forced to perceive themselves and their sexuality according to the wants of men, "keeping women's eyes lowered to their own bodies, glancing up only to check their reflections in the eyes of men" (155). In her chapter "The Sex-Seekers," Friedan makes a similar point: "What happens when a woman bases her whole identity on her sexual role; when sex is necessary to make her 'feel alive'?" (265). Hence, even though Mildred understands her desire, she can only act when Juan assures her *he* desires *her.*

Nevertheless, Juan and Mildred have selected each other for this act, go through some absurd theatre, and then fulfill their desires in one of Steinbeck's typical settings—a barn next to an abandoned farmhouse given up to weeds. Afterward, the first thing Mildred does is have Juan hold her small mirror for her while she puts on makeup. She oscillates between the truth and lies—asking what Juan really thinks of her, if they will go to Mexico together, why he really left the bus. To her credit, she prefers the truth. Still, it is hard for her to break out of the confines of the norm. She needs to know that Juan thinks she's more than "all right": "Do you want me to lie?," Juan asks; she answers, "I guess I do a little. No, I don't" (196). Mildred knows the whole little affair is wrapped in fantasy and romance, and finally she separates herself cleanly and comes away unaffected: "I wish it could go on a little more . . . but I know it can't. Good-by, Juan" (198). The old stallion and the young filly meet in a barn, both of them using each other and knowing it and both satisfied by the result; Juan has the experience, Mildred is learning. As she leaves, still free and in a sense still becoming, hers is the more enviable situation. Juan goes back to his bus.

For sheer sexual power, Camille Oaks (she chose this name for the trip from an advertisement featuring a poster girl) is the strongest character in the novel. Ironically, she is a victim of both male competition and the beauty myth, but this fact does not reflect as badly upon

Camille as on the men who fawn over her. As Gonzales and Gladstein have already observed, "Steinbeck's explication of Camille is a case study of Darwinian naturalism" (162). Camille herself, other women, and even her male doctor (after he quells his own desire and acts like a professional) recognize that her sensual body exudes sex. Men react to her as if they unconsciously sense pheromones at her approach. Gonzales and Gladstein note that although "Camille is beautiful, sexually desirable, and intelligent, all of the men in the book treat her as nothing more than a sop for their sex drives" (163). Her relationships with men are marred by the intensity of the sexual selection process her desirability initiates: "Men fought each other viciously when she was about. They fought like terriers" (*Bus* 74).

Still, Camille exercises her power within the limits set for her. Louie, the Greyhound bus driver, is the mouthpiece for all misogynists, and clearly Steinbeck does not portray him as a hero. His character is established immediately, when he cheats George, a black swamper, out of a reward for a lost wallet. A petty man, Louie thinks of himself as a stud who can easily seduce women, whom he calls "pigs" (67). Louie tries all of his pathetic wiles on Camille, but she knows by experience every move he will make. He tries to unsettle her with stares, but her eyes match his—she will not look meekly away. Juan also assesses her but quickly realizes that he is too old and that she is more than his equal; Camille is too self-possessed to be affected by him and he knows it.

Her attempt to teach Norma a few things about makeup and hairstyling demonstrates that Camille is adept at ornamentation. Her exaggeration of those attributes that culture deems beautiful is reminiscent of Darwin's observation of the pains primitive tribes take to augment physical attractiveness by conforming to a standard. Darwin also notes that among the civilized cultures external appearance plays a great part in sexual selection (just as it does among other animals). He writes: "The general truth . . . [is] that man admires and often tries to exaggerate whatever characters nature may have given" (*Descent*, Modern Library 889). Among primitive peoples such attempts at adornment involve mutilations (piercings, scars, etc.), and of advanced cultures he notes, "In the fashions of our own dress we see exactly the same principle and the same desire to carry every point to an extreme" (889). As Wolf points out, the recent imperatives of the beauty myth mean women today must suffer starvation, liposuction, acid skin peels, and cosmetic surgery to achieve Darwin's "extreme." In an absurd example of this reality, Steinbeck depicts Louie trying to set a fashion trend by growing the nail on his little finger to an extraordinary length. Camille's prowess with the game of adornment, to which she has given in (for a time she tried severe dress to put men off, but it failed), demonstrates Steinbeck's awareness of an oppressive cultural standard.

Camille has decided to take what advantage she can of her position and performs at stag parties for easy money: "[S]he made fifty dollars for taking off her clothes and that was better than having them torn off in an office" (74). She is a victim of a society lost in advertising, worshipping the image of the poster girl. Camille must be hounded and abused by men, for she happens to look like the exaggerated calendar women hanging in the diner that depict "improbable girls with pumped-up breasts and no hips—blondes, brunettes and redheads, but always with this bust development, so that a visitor of another species might judge from the preoccupation of artist and audience that the seat of procreation lay in the mammaries" (3).

As he shows in such works as *The Winter of Our Discontent* and *America and Americans,* Steinbeck despises the whole advertising industry; his friend Webster Street said, "I think he hated dishonesty of any kind, you know—hypocrisy, particularly things like advertising and campaigning and that kind of thing" (122). Certainly his sketch of the self-destructive, failed advertising man that comes between Camille and her friend Loraine confirms his dim view of the profession. That an advertising man should come between Camille and Loraine is significant, for the advertising men have created the mold (what Wolf refers to as the "Iron Maiden" [17]) that this hapless woman fits, and they have created the desire for her by men amazed to see the standard of adornment alive in their midst.

Camille wants to lead a normal life, perhaps to raise a family—an ideal conceived in her imagination, ironically, from "advertising in the women's magazines" (74). She is clearly a victim, but to her credit she is not weak; she is a survivor who can deal with the stupid, lust-crazed men who would ruin her life. No man in *The Wayward Bus* has a prayer of dominating Camille, not Louie, Juan, the fumbling Elliot Pritchard, nor the pathetic Pimples. By saying either yes or no, she chooses what men she will have. Still, "there was a great deal of sadness in her" (74) because she is alone, a forced practitioner in a system not of her design; nevertheless, by knowing the system well and by being tough, she has achieved a measure of control over her own sexuality.

In fact, while Camille is a Darwinian creation, her plight perfectly illustrates the oppressive power of the beauty myth. Although many women cannot meet the standard of the Iron Maiden, Wolf observes that life is not any easier for "beautiful" women. Such a woman's body is "defined and diminished by pornography," which "inhibits in her something she needs to have, and gives her the ultimate anaphrodisiac: the self-critical sexual gaze" (149). Camille is alone because other women resent the perfection of her body; as Wolf writes, "beauty thinking urges women to approach one another as possible adversaries until they know they are friends" (75). Camille knows men fight over her and women do not like her, but "there wasn't anything she could do about it" (*Bus* 74). As a walking magazine picture, she has no friends, often is a victim of attempted rape or violence, and finds escape in domestic fantasies she has seen in women's periodicals.

Norma, being rather plain in physical form and personality, lacks the dynamism of Juan, Mildred, or Camille. Her experience is very limited, and in her imaginary love affair with Clark Gable she has given herself over to Hollywood's illusion factories. But Norma is not a fool; she knows Alice rummages through her things and knows how to keep Pimples at bay. Her plan, wisely, is to leave Rebel Corners. Although she has a hopeless desire to be discovered as a star, she is tough enough to survive: "Her high, long-legged dreams were one thing, but she could take care of herself too" (62). She shows her resolve when she decides to part company with Alice and Rebel Corners.

Under Camille's tutelage, Norma spruces up, to conform more expertly to the standard of physical attractiveness. Camille knows that makeup corresponds to a woman's self-esteem: "It would be fun and company to make this girl [Norma] over and to give her some confidence" (97). Advertising and the Iron Maiden image make it very difficult for a plain woman like Norma to have much confidence; as Wolf writes, "Advertising aimed at women works by lowering our self-esteem" (276).

Norma's makeover does give her confidence, but it also arouses the interest of Pimples, who, along with everyone else in the novel, loses his inhibitions once the bus has become stuck in the mud. As Gonzales and Gladstein observe, Pimples tries to take advantage of Norma's sympathy for his ugliness (162) and forces himself on her in the back of the stalled bus, a very different scene from the mutually responsive affair between Juan and Mildred. Norma first tries to reason with

Pimples, telling him that someone might come and see them, but finally "her work-hardened muscles set rigidly," and she beats him off with her fists (*Bus* 204). Here, sexual selection operates both ways; the undesirable Pimples may seize his woman with wolflike rapacity, only to find that she is not his and never will be. Norma exhibits compassion—always a significant trait to Steinbeck—while Pimples proves himself to be an animal, out of control of his sexual impulses and in many ways beneath contempt.

Elliot and Bernice Pritchard are obviously loathed by the author; they represent the middle-class respectability and hypocrisy that Steinbeck made a career of attacking. They even have earned the contempt of their daughter, Mildred, a character Steinbeck appears to hold in higher regard. Bernice and Elliot Prichard live in an artificial world and, unlike Juan or Mildred, can neither recognize a natural impulse nor act upon it.

Elliot, the president of a large company, thinks of himself as a leader, tough like his ancestors. But, foreshadowing Steinbeck's comments in *America and Americans* about our atrophy and waning survival drive, Pritchard proves himself such an unnatural creature that he could not hope to survive outside the rarefied atmosphere of corporate America. Ernest Horton, another stranded passenger, points out that in a real emergency Elliot could not fix a car nor even procure food for himself, facts that lead Pritchard into a daydream of denial, in which he sees himself going out to bring down—of all things—a cow to supply meat for the group. Later, he tries to pick up Camille, half believing his own lie that he wants to hire her as a receptionist, and falters when she holds up the truth to him—that he's looking for a mistress. She also confronts him with his sleazy lust, his staring at her naked body during a performance for The Octagon International: "I don't know what you get out of it and I don't want to know" (195). Camille reveals the beast in Elliot, and he cannot stand the truth; she makes him feel "naked" (194).

Elliot's wife, Bernice, also denies reality and tries desperately to make her marriage appear perfect. A physiological problem that prevents her from enjoying sex has slowly led her to manipulate her weak husband, strangling his natural desire. Gonzales and Gladstein write that they are "sure Steinbeck means for readers to be pleased when Elliot Pritchard, driven by his interchange with Camille to a recognition of his sexual needs, walks into the cave where his wife lay, and brutally asserts his marital rights" (165). However, this scene is perhaps one of the most pathetic of Steinbeck's many depictions of characters who drop the artificial masks of civilization and entirely lose control of themselves. Elliot's rape of his wife, enacted in the cave—another of Steinbeck's primitive arenas—demonstrates the wretched state of people who live a lie, denying or withholding natural desire until they are driven to acts of desperation. Once Elliot stalks off, Bernice mutilates herself in an act of animal fury—rubbing dirt into the bloody scratches she has made on her face.

With the exception of Horton, a man of scrupulous honesty who is unaffected by the sexual atmosphere of the novel, all the characters of *The Wayward Bus* are to varying degrees subject to or victimized by biological drives. The lower echelon are unaware of but enslaved by their animal natures.

Alice Chicoy suffers on the wrong end of the spectrum of sexual selection; she feels ugly, past her prime, and fears that she will lose Juan to a more desirable or youthful female. Like Norma, her confidence is directly tied to her lack of beauty. As a result, she hates herself and fears any woman who fits the beauty ideal, such as Camille. "Older women fear young ones," Wolf writes, and "young women fear old, and the beauty myth truncates for all the female life span" (14). If some kind of sexual selection does exist in human relationships, the beauty myth perverts it.

With her self-loathing, Alice is happy to have some time alone once the bus pulls away from the diner. Gonzales and Gladstein suggest that Steinbeck devotes chapter 11 to Alice merely because he "is bent on reducing this woman to her most vile state" (169). Yet Alice is not singled out; soon Pimples, Elliot, and Bernice are seen in their most vile states as well. One good reason for chapter 11 is that, alone, Alice must face the truth about herself, that she is getting older and is afraid of her mortality. We also see her try to use alcohol to hide from these realities. The lonely lunchroom is for Alice what the cave is for the Pritchards or what the back of the bus is for Pimples—the place for veneers to fall away, for the true self to emerge.

Alice tries to dwell in denial during her day alone, but even alcohol and imaginary drinking buddies cannot stave off the horrifying memory of her dying mother (116). She skips from this to her own physical aging and tries desperately to defy the truth: "'I'm getting along— That's a god-damned lie,' she shouted, 'I'm as good as I ever was'" (117). She plunges into more Old Grandad, and then tries cosmetics to put some youth back into her face. Even more drunk, she imagines an affair which dissolves into the memory of Bud, who brutally robbed her of her virginity, and she shouts at herself in the mirror, "None of it's any damn good" (119). We have a horrifying glimpse of her life at Rebel Corners: a life in an isolated lunchroom, married to a man with whom she must "walk on eggs all the time to keep . . . happy" (119). She sees herself as a "dirty drunken old bag" (120).

The smallness of her circumscribed world, the limits which she feels so painfully, all are magnified by her drunken pursuit of a housefly. Alice finally passes out, which is what she wanted all along, and a fly dips "his flat proboscis into the sweet, sticky wine" (122). Gonzales and Gladstein suggest that Steinbeck uses this "sassy fly" to thumb his nose at all wives; if so, what a poor symbol in which to embody male *machismo*! This fly can hardly be seen in a positive light, especially when flies have been linked so carefully to Pimples, easily the most disgusting character in the novel.

If Steinbeck had left Alice without including this chapter, we could easily have written her off as a hateful, one-dimensional character. However, chapter 11 shows us her desperation and loneliness and the absolute limitations that hem her in. We see that living with Juan, this man who beats her and makes her tiptoe about in fear, is not a dream existence, and her self-loathing, born of an obsessive focus on aging and ugliness, confirms the oppression of the beauty myth. Perhaps most important, Alice's plight further illuminates the theme of denial or repression of the biological realities of living in a body, an aging and imperfect shell. Alice cannot face old age and waning desirability any better than Elliot can face his repressed sexual desire, Bernice her lack of desire, or Pimples his inability to succeed sexually. Neither alcohol, cosmetics, nor money can hide these facts, especially when one is alone or suddenly cast out into the natural world.

Two male characters represent the lowest of the sexual/biological scale in *The Wayward Bus*. Pimples is the fly, one of Steinbeck's grotesques, a man who tries to satisfy his cravings—be they for a woman or for a custard cream pie—with the delicacy of an insect. Van Brunt, an elderly passenger, is like Alice; he feels his years and fears the helplessness that age brought, in this case, to his father. Silent cerebral strokes fuel his anger. Feeling his gathering weakness, like Alice, he lashes out at others with hatred. His illness has robbed him of the ability to control his emotions, including sexual desire: "The stroke had knocked the cap off one set of his inhibitions. . . . He was pantingly drawn toward young women, even little girls" (200). Both Pimples and Van Brunt are ugly, arguably uglier than Alice, and the two men are completely enslaved to their bodies. Pimples, hardly above a fly, follows his appetite or hormones; Van Brunt is a victim of his own clogged arteries and dying brain cells.

The Wayward Bus, as harsh a view of *Homo sapiens* as *In Dubious Battle,* fully extends Steinbeck's biological perspective into human sexual relations. The characters are coldly measured by their control over biological reality. The novel is a grotesque portrayal of sexual selection, made all the more distressing because of the repression and perversion of sex by a warped society. Juan and Mildred meet in a barn, satiate their needs, and play at romance until they both know the game is over. Camille Oaks sees the sexual power plays of society for what they are and understands the role her body and the men around her have forced her to play. Norma sinks into the romantic bathos that Hollywood has made of love, but she is tough enough to face grotesque reality when it reaches out to grab her. Elliot and Bernice Pritchard wrap their lives in the lies of middle-class respectability, and when the truth of their own sexual power game is exposed in the cave, they scurry to deny what they see. Alice and Van Brunt live in a waning physical world; as their bodies run down and fear possesses them, they realize the smallness of their lives. Pimples is disfigured, undesirable, and so base in his nature that he is little better than the fly Steinbeck has made him. *The Wayward Bus* is meant to be a slap in the face, but certainly not just for women. Rather, it attacks the inherent weakness of a society that has gone Hollywood and tries to cover up biological reality with makeup, booze, or a handsome business suit.

❂ ❂ ❂

Darwin and Steinbeck approach human sex as they do most other activities of *Homo sapiens* and ignore the poetic lore heaped upon this biological act. Both see the animal nature of human love and try to understand it; as often happens, they come to similar conclusions, both of them recognizing the dynamics of sexual selection. Certainly they share a limited biological conception of men and women, tending toward some reprehensible generalizations: man is selfish but, because of the demands of sexual selection, also the greater achiever; woman is selfless, the mother vessel of the species, but more limited in the range of possibilities. Yet Steinbeck does seem to realize the limitations of these roles and many of his female characters come to us as victims of their circumscribed sphere. Some, like Camille Oaks or Alice Chicoy, feel imprisoned by the narrow scope of their lives in a patriarchal society. Significantly, however, Steinbeck could never conceive of a woman who finds an alternative, who blazes a new path. Such a woman is not part of his or Darwin's biological conception of the species. Here, their scientific search for truth lets them down.

5

From *Homo sapiens* to Human—
Evolution of a Hero

"What saved Steinbeck from constant excess was a compassion that was, in much of his writing, balanced and disciplined by a very objective view of the world and of man."
—*Jackson J. Benson,* The Short Novels of John Steinbeck, *1990*

So we are animals, part of the same whole that is the aggregate of all life on Earth. But surely we are something more, and isolating that "more" has been the business of theologians and philosophers ever since we evolved enough to consider the problem. Darwin and Steinbeck, from their scientific view of the human as a species, conclude what the difference is, or may some day be. The problem involves nothing less than defining humanity. Strikers and soldiers may fight with the brutality of beasts, men and women may love with the ferocity of animals, but there are also those with a wider vision. Monkeys never negotiated a peace, designed a machine, nor estimated the size of a galaxy. This chapter examines how Darwin and Steinbeck extract the "more," the human, from the "less" in *Homo sapiens.*

We can trace in Steinbeck's work a quest to find what qualities in a human are different from those of a beast. In this sense, Steinbeck's fiction follows a line of development, a kind of evolution in itself, in which the author creates a hero, exposes a flaw, and tears the hero down. Eventually he arrives at a hero who resists destruction, who proves a worthy example of what, in Steinbeck's view, a good human being should be. As Steinbeck and Darwin follow the same course, how they define the human in humanity is remarkably similar. Indeed, the "Doc" characters and Charles Darwin share many traits. This bright view of humanity came gradually for Steinbeck, a product of the kind of developing process the novelist describes in his response to *Steinbeck and His Critics:* "It is always astonishing to read a critique of one's work. In my own case, it didn't come out that way but emerged little by little, staggering and struggling . . . then, after the fact—long after—a pattern is discernible, a clear and fairly consistent pattern" ("Postscript" 307).

Before defining Steinbeck's conception of the hero, or the humane person, we should examine how Darwin defined a person at the highest level of civilized development. The naturalist's writing, particularly his *Autobiography, The Descent of Man,* and various letters, presents a reasonable picture of those qualities Darwin would have found of the highest order in humanity. Not surprisingly, these characteristics make up much of Darwin's own nature as he and others perceived it.

Darwin wants to see things in a most unprejudiced way, to observe without preconception, and to do so requires a mixture of bravery and humility. Bravery asserts a new idea that runs counter to most popular scientific and theological concepts; humility presents it in the truest way, without distortions caused by an eye toward fame. Observation and inductive thinking are crucial to the naturalist's method. As he writes in his autobiography, "I worked on true Baconian principles" (*Autobiography* 42). After noting that one of his best abilities is his power of observation, he explains his desire "to understand or explain whatever I observed" but to do so without following "blindly the lead of other men" (55). He also writes, "I have steadily endeavored to keep my mind free so as to give up any hypothesis, however much beloved (and I cannot resist forming one on every subject), as soon as facts are shown to oppose it" (55). In this section Darwin adds that he has been led "to distrust greatly, deductive reasoning in the mixed sciences" (56). This process of thought leads him to a wider view, as suggested in an article by T. H. Huxley (one of Darwin's avid defenders): "The teleology which supposes that the eye, such as we see it in man, or one of the higher vertebrata, was made with the precise structure it exhibits, for the purpose of enabling the animal which possesses it to see, has undoubtedly received its death-blow. Nevertheless, it is necessary to remember that there is a wider teleology which is not touched by the doctrine of Evolution, but is actually based upon the fundamental proposition of Evolution" (*Autobiography* 316). Darwin recognizes the folly of a narrow view in an 1860 letter to Sir Charles Lyell (the geologist who greatly influenced the naturalist): "It is funny how each man draws his own imaginary line at which to halt" (*Autobiography* 241).

In *The Descent of Man,* Darwin associates the narrow, nonholistic view with the primitive mind: "He who is not content to look, like a savage, at the phenomena of nature as disconnected, cannot any longer believe that man is the work of a separate act of creation" (Modern Library 909). (Not only "savages" are chastised for a narrow view; after meeting Thomas Carlyle, Darwin makes a rare critical judgment, disliking the philosopher's sneering ways and observing that "his mind seemed to me a very narrow one" [*Autobiography* 39]). An enthusiastic comment in an 1859 letter from a man who has embraced Darwin's theory is especially interesting, in light of the ideas of Steinbeck and Ricketts: "I must give up much that I have believed and written . . . Let God be true, and every man a liar! Let us know what *is*" (*Autobiography* 241). Charles Darwin uses and inspires others to use an inductive method of thinking based on observation that avoids prejudice, takes in the whole, and finds truth by an unfettered acceptance of what is.

Steinbeck believed in the inductive method. In a foreword he wrote for *Between Pacific Tides* (a book published by Edward F. Ricketts and Jack Calvin in 1939), Steinbeck discusses cycles of human thought and how each period in history warps that generation's view of the truth. "Modern science, or the method of Roger Bacon," he writes, "has attempted by measuring and rechecking to admit as little warp as possible, but still some warp must be there" ("Foreword" v). Admitting the scientific method's flaw as a time-bound approach, Steinbeck still applauds it since "one can indulge perhaps the greatest human excitement: that of observation to speculation to hypothesis . . . a creative process, probably the highest and most satisfactory we know" (vi). For Steinbeck, this creative process is the most important form of human thought but must nevertheless be tempered by the humble recognition that its products will not be free from "warp."

The inductive method of thinking led to Darwin's greatness, which was recognized in his lifetime but is not a product of some conscious desire to become famous. Darwin searched for truth in nature, and fame was a by-product rather than a goal: "I did not care much about the general public. I do not mean to say that a favourable review or a large sale of my books did not please me greatly, but the pleasure was a fleeting one, and I am sure that I have never turned one inch out of my course to gain fame" (*Autobiography* 32). In fact, Darwin tried to duck fame and controversy; he wanted to get on quietly with his work and probably knew that any inflation of ego would blur his ability to observe and think inductively. As we know from Benson's biography of Steinbeck, *The True Adventures of John Steinbeck, Writer,* and the author's own comments in his *Grapes of Wrath* journal, (*Working Days*), Steinbeck also dreaded fame. He even feared that winning the Nobel Prize might hurt his work (which it did).

The personal humility of both men can clearly be seen by the way they accepted the greatest honors ever presented to them. In 1864 Darwin received England's highest scientific honor: the Copley Medal of the Royal Society. As his son Francis explains, Darwin felt too ill to accept the award personally, and the naturalist's response to the honor illustrates his view of fame: "The Copley, being open to all sciences and all the world, is reckoned a great honour; but excepting from several kind letters, such things make little difference to me" (*Autobiography* 274). Benson has carefully recorded Steinbeck's reaction to receiving the Nobel Prize; the author was ecstatic, but at a press conference his humility helped feed critics who were hungry to deride him. Benson describes what happened after a reporter asked Steinbeck if he really deserved the prize: "If he had a little more ego, he would have lost his temper; if he had been more of a politician, he would have said that was for the committee to decide; but being John Steinbeck, he looked straight into the eyes of the reporter and said, 'Frankly, no'" (*True Adventures* 915).

Charles Darwin praised observation and induction and could submerge ego enough to accept what appeared to him as truth. The "is" thinking of Steinbeck and Ricketts works in precisely the same way. To each of them, this cognitive process, requiring absolute avoidance of prejudice in order to be most successful, represents a superior way of ascertaining truth. Although such a way of thinking represents a more advanced mind, both Darwin and Steinbeck focus on a single quality that separates the beast from the humane person: sympathy.

The naturalist reveres sympathy and estimates other individuals and humanity in general by this quality; he finds its cousins, empathy, compassion, and cooperation, to be essential components of the highly developed human being. Writing of Lyell in his autobiography, he lauds him with the observation that "one of his chief characteristics was his sympathy with the work of others" (*Autobiography* 33). In recounting his winning of the Royal Medal in 1853, Darwin typically plays down the award but seizes on the warmth of a congratulatory letter from J. D. Hooker (another Darwin defender): "Believe me, I shall not soon forget the pleasure of your letter. Such hearty, affectionate sympathy is worth more than all the medals that ever were or will be coined" (*Autobiography* 172). Writing of his father, whom Darwin admired highly, he calls him "the most perfect sympathiser" (*Autobiography* 91).

His own admiration for this quality finds its way into *The Descent of Man*. Sympathy is chief among the "moral qualities" which distinguish humans from animals. Morality arises from social instincts, which we share with other animals. "These instincts are highly complex, and in the case of the lower animals give special tendencies towards certain definite actions," Darwin writes, "but the more important elements are love, and the distinct emotion of sympathy" (Modern Library 912). He also notes that these social instincts "are highly beneficial to the species," as mutual cooperation facilitates survival, and "have in all probability been acquired through natural selection" (912). Further, he calls sympathy "one of the most important elements of the social instincts" (913), crucial to the survival of a species as it develops cooperation. In the last line of the book, he writes that "sympathy which feels for the most debased" is foremost among the human's "noble qualities" (920). Darwin uses sympathy as Steinbeck uses the term *simpatico* in *The Log*: it promotes active cooperation. Both Darwin and Steinbeck might naturally regard sympathy as important for it requires one to see and feel beyond the self and is an emotion that works in tandem with their method of thinking.

Sympathetic emotions and inductive reasoning will likely defy prejudice and ultimately bring one closer to truth. For Darwin the goal is to define what is true; for Steinbeck the goal is to depict the truth. The whole notion of "what actually 'is'" involves moving beyond personal and cultural preconceptions. An inductive thinker who observes without bias, submerges ego, and possesses sympathy stands the best chance of knowing truth. Above all else, the work of Darwin and Steinbeck represents an effort to present, with the aid of science, a truthful view of our place in the world—no matter whose prejudices might be destroyed as a consequence.

The qualities that Darwin most admires define Steinbeck's hero, the kind of human being that the author spent nearly a lifetime searching for and portraying. The most noble of Steinbeck's characters think inductively, accept the "is" about themselves and the world, and extend beyond self to become the quintessence of sympathy, compassion, and cooperation. These people, exceptionally honest with themselves and others, are the ones who will resist the beast and push humanity another step forward. In times of struggle, they are the cooperators, and they are the ones who will endure.

FROM CUP OF GOLD TO THE RED PONY

Steinbeck's first major protagonist, Henry Morgan of *Cup of Gold*, has the potential of greatness but falls short. Robert Morgan, Henry's father, knows (as Darwin once mused about people in general) that a mind has certain boundaries which cannot be exceeded. His son, young and open, eagerly pursues his visions; as Robert exclaims to his wife, "He tests his dreams, Mother, and I—God help me!—am afraid to" (12). Henry possesses the courage and strength to leave the valley and take on the world. A keen observer, he picks up the sailor's ways or the plantation owner's business in quick order, and at The Three Dogs inn he studies the collection of nationalities intensely. He comes to Barbados to serve James Flower, a plantation owner, but soon exceeds the older man's abilities because Flower, for all of his extensive knowledge, cannot conceive a single original idea; due to this want of imagination, he cannot advance his learning, nor act upon it. Like Robert Morgan, Flower lacks vision. He cannot even verify a theory, for he does not have "the spirit of induction" and therefore can form "no design of the whole" (56). Henry, however, possesses these powers in some degree and so, once learning what Flower knows, he can pursue lofty dreams.

But despite such strength of mind coupled with his physical strength, Henry's flaws eventually defeat him. His view of the whole is limited; lacking sympathy, Henry cannot see beyond his own needs and becomes enslaved by his dreams: "He was merciless" (62). On the plantation he becomes master, for he knows precisely how best to intimidate the slaves into nearly working themselves to death for him. As he misuses slaves, so he misuses women as objects to satisfy his desire. He settles on Paulette, whom he looks upon as "a delicate machine perfectly made for pleasure, a sexual contraption" (65). She comes to love him very much, but to him she remains simply a "contraption," and when she displeases him, he threatens to have her whipped.

Finally, the coldness of his rational mind is overtaken by the focus on his own wants, in the form of a beautiful woman, La Santa Roja, and "[h]e struggled madly against the folding meshes of his dream" (98). As Merlin, an old hermit poet, tells Robert, "People have so often been hurt and trapped and tortured by ideas and contraptions which they did not understand" (108). La Santa Roja does not hurt him, but rather it is Henry's dream of what she may be that eventually tortures him.

Henry has the strength to reach his dream, and he does so in a typically Darwinian manner: he leads a group of hungry, toughened men in an effort to take over Panama from the fat and lazy Spanish. Henry forms a "race of pirates": "Out of his mob of ragamuffin heroes he wanted to make a strong, durable nation, a new, aggressive nation in America" (87). These are not the kind to pay only lip service to survival, the opposite of the rulers of Panama who sit idly upon their "Cup of Gold." The Spaniards "had grown soft in their security," because for too long they had not had to fight for the "impregnable" city (114), and they quickly fall to Henry Morgan and his ferocious, stampeding pirate race.

Confronted at last by La Santa Roja, Henry's dream in the flesh, his failure quickly becomes apparent. Physically, the Red Saint—now simply Ysobel—is beautiful, but not in the way Henry had dreamed: "He was staggered at such a revolt against his preconceptions" (141). He tells her his dream of her, that they burn for each other, that they must be married. He tries to make his vision work, but she is the realist, telling him they do not burn for each other and, further, that his dream is not even original: "I am tired of these words that never change" (143). She wants a forceful, brutal realist but instead finds him merely a "bungling romancer" (144). She challenges him with "truthfulness" (145).

Henry begins to be tortured for he sees his accomplishments from a wider perspective. He knows he has led men through the jungle, killed hundreds, and burned a city for a vision that does not exist; with this knowledge, he sees the conquest as "after all, a pitifully small and circumscribed destruction" (147). He is a cruel master of men, but nothing more, and his dream is a hollow one. Henry sits alone in the Hall of Audience in Panama, and "his eyes, those peering eyes which had looked out over a living horizon, were turned inward" (158).

Henry cannot reach beyond the borders of his own ego, and finding a hollowness within himself, the bold pirate shell collapses. Ysobel points out that "the boastful child" who "thought his mockery shook the throne of God" has died; Henry agrees (162). In the end, he lives for security and nothing more, working for Charles II as lieutenant governor of Port Royal and dispassionately hanging former pirate friends. Henry Morgan dies a petty official, as lazy and stupid as the kind of man he had so easily beaten in Panama. *Cup of Gold* portrays a man of strength and vision whose dishonesty and selfishness allow his egocentric dreams to swell until they burst, leaving a broken old man wondering whatever became of his greatness.

Joseph Wayne, in *To a God Unknown,* comes closer to knowing the whole, in fact passionately desires to know it, but fails because he cannot accept what "is." Louis Owens accurately describes Joseph Wayne as a man who sees only parts of the whole, whose religious connection to the land locks him into a cause-and-effect teleology (18, 20) (that is, if he worships hard enough he can make something happen in nature). Like Henry Morgan, Joseph Wayne fails to see outside of his own intense vision; unlike Morgan, however, his failure does not rest on sheer selfishness. He leaves his father to satisfy a hunger, but not for adventure, gold, or greatness. Instead, he looks for land to call his own and to allow his brother Benjy to have more of the family farm in Vermont. Reaching Nuestra Senora in California, Joseph finds the land he has hungered for, and becomes so enamored with it that he loves it physically—flinging himself upon it, face down, while "his thighs beat heavily on the earth" (11). He has a "hot desire" to make contact with the land, and his desperate lovemaking foreshadows the kind of monomania that led to Henry Morgan's undoing.

Joseph's obsession with the land increases when he hears of his father's death, and he comes to believe that the elder Wayne's spirit has entered a great old tree on the farm. He begins to converse with it, stopping only when Juanito, a ranch hand, comes near. Just as Henry Morgan at first senses danger in his obsession with La Santa Roja, Wayne feels a bit uneasy about his feelings for the tree and the land: "Joseph wondered why he did not try to escape from the power that was seizing upon him" (27). At first his mind is free enough to see outside his vision: "What I feel or think can kill no ghosts nor gods" (27). Soon, however, Joseph's mind narrows upon the idea that he has become one with the land, perhaps even its master. "All things about him, the soil, the cattle and the people were fertile, and Joseph was the source, the root of their fertility," Steinbeck writes. "[H]is was the motivating lust" (34). Joseph's animism is not a sin, as his brother Burton

—entirely caught up in the preconceptions of his Christianity—would have Joseph believe. Rather, Joseph's sin is that he creates his own preconceptions, he mythologizes the land, worships it, and thinks that he wields power over it. He projects his own hopes and beliefs onto the land, seeing his father in the tree, life in the glade, and power in the great rock in the glade's center.

Joseph looks too hard for symbols, and when he cannot find them he creates them. Just after asking Elizabeth to marry him, he notices that the night is "too unimpressed," and he throws off his hat, "but this was not enough," so he whips his own leg viciously (52). At his wedding, Joseph refuses the traditional symbols and rituals of the church, which he considers "a doddering kind of devil worship" (72). The tolling bells in the belfry mean God to Joseph: "Here's God come late to the wedding. . . . This ties in. . . . This is my own thing and I know it" (72). To Joseph, the real wedding does not occur in the church at all, but when he and Elizabeth go through the pass and into the valley to his land. On the way, a cloud rests on a ridge and he sees it as a goat's head. He believes in the mythological sign he has just created: "If I will admit the goat is there, it will be there. And I will have made it" (86). When he hears of Benjy's death, he believes he can have no pure, individual feeling, for "all things are one, and all a part of me" (92). His is an attempt at a holistic outlook, but with a fatal flaw: seeing himself at the center, his perspective lacks the inductive/humble quality of Darwin's or Steinbeck's thinking. Others, such as Rama, the strong housewoman who worships him as a "godling," only foster his belief that he is the most important creature in the whole (99).

Joseph fears above all else a killing drought and knows only too well that the land goes through dry seasons. Waiting for rain (his brother, Thomas, tells him, ironically, "You may keep the rain away if you're too anxious" [118]), Joseph dabs pig's blood on the tree, and the ritual works this time—rain comes. The following year, however, Joseph's wishes and nature's course diverge. After Burton kills the tree by girdling it, Joseph feels he has no counsel and becomes fearful as drought begins to move over the land. Elizabeth, who strives to be a rational thinker, comes to think like Joseph during her pregnancy, but after giving birth feels normal again: "I was oversensitive. . . . When I was carrying the child, little things grew huge" (181). She tells him that she went to the rock in the glade, loved it as he has, but now has become afraid for she sensed evil there. Elizabeth goes with Joseph to the rock to insult it for frightening her, and while climbing over the green moss she slips and breaks her neck. Joseph cannot comprehend how her life could be snuffed out so quickly and "without any meaning" (190). "He wanted to make himself know what happened," Steinbeck writes (190).

Unable to accept the accident and lacking the counsel of the tree, Joseph tries to find new solace by believing that Elizabeth's spirit has gone into the rock. Leaving her body to Thomas, Joseph goes to a pool, now stagnant because of the drought, and comes face to face with the cruelty of nature. When he realizes the drought cycle has come on, Joseph feels no awareness of spirits, no control, and he witnesses first-hand the struggle for survival. Wild pigs catch eels from the receding pond, only to fall prey to a mountain lion. Joseph cannot accept this picture and wishes he could shoot the lion. He no longer comprehends nature as clearly as when he first came to the land and understood the uselessness of killing a boar that ate its own children, knowing the old male would eventually produce more offspring than it killed.

The predicament of Joseph Wayne, and humanity in general, is underscored by a passage immediately following his visit to the dying pond. The omniscient narrator presents the view from the "brain of the world," the greater nature of earth, and personifies this larger entity, which says, "I will endure even a little discomfort to preserve this order which has come to exist by accident" (196). Joseph's feeling that he has built fertility into the land and has some divine place in it is undermined by this passage, introduced and closed by the phrase "size changed," which has the effect of widening the aperture to reveal that Joseph's personal religion, like all religions, has no effect on the greater natural cycles which human and earth alike must endure.

As the drought takes hold, Joseph thinks, "The duty of keeping life in my land is beyond my power" (206). Yet he cannot leave the ranch, even as it dies, for he fosters a hope that it will come alive again in the following year. Although he gives his child up to Rama as a sacrifice, saying, "It seems to me a thing that might help the land," nothing improves (226). As people leave him, the animals die off, and the farm becomes deserted, Joseph Wayne's sphere—like his mind—narrows and narrows. Finally he is alone in the glade, with the rock that has become his final object of worship, telling his friend Juanito, "Only this rock and I remain. I am the land" (240). He tries to sacrifice a calf, letting its blood run into the ground, but to his obsessed mind there is too little blood to do any good. Accidentally cutting his wrist, he climbs on the rock and lets his own blood run out—he becomes the final sacrifice.

The fact that rain comes immediately after the sacrifice shows that Steinbeck walks on the edge of spirit and reality. The rain begs the question, did the tree and the rock really mean something outside of Joseph Wayne and did the sacrifice work? Several critics, such as Richard Astro (*Steinbeck and Ricketts*) and Robert DeMott ("To a God Unknown"), read Joseph's sacrifice as positive—he has given himself up for the salvation of the whole. Indeed, the timing of the rainfall underscores this possibility. However, the omniscient voice of the earth brain suggests another explanation—that the rain simply came at the end of the cycle.

In the early drafts of the novel, Steinbeck considered Andy Wayne —forerunner of Joseph—to be a psychological case history, as Benson notes (*True Adventures* 172). Benson quotes a letter from Steinbeck which clearly indicates that the author viewed Andy Wayne from a psychological and anthropological perspective (173). In the finished novel, Joseph Wayne still seems a case study, not a demigod. The book ends with Father Angelo's ironic observation that with the drought broken, Joseph must be very happy. The reader knows, however, that he has died and might accept the idea that his sacrifice brought on the rain, but Steinbeck holds out the tragic possibility that a completely deluded Joseph Wayne killed himself just before the natural cycle would have swung back in his favor.

Clearly, Joseph Wayne, in striving to understand the whole, has miscalculated his place in it. Joseph's fatal error is that he projects himself onto the land so intensely that he cannot accept the inevitable natural cycle and his subjection to it. Giving his blood to a rock is as useless a gesture as passionately beating his thighs against the earth. Lost in his own preconceptions, Joseph Wayne fails to break through to true understanding.

While Morgan and Wayne cannot see through to what is, Junius Maltby, in chapter 6 of *The Pastures of Heaven*, can see the wide vision and accept it. Yet Maltby acquiesces all too easily, without the strength, passion, and drive of a Morgan or a Wayne; Maltby is simply too weak to make his long vision survive in the world. Perhaps, in writing the 1932 book with Morgan far in his wake and Wayne still a problem to be worked out, Steinbeck created a character of opposite disposition to his other two tall protagonists to see how a humbler man would stand up.

Easy to admire, Maltby does not fall prey to a vision of his own making. After he marries a widow, Mrs. Quaker, he revels in laziness, letting her farm go fallow because "he liked the valley and the farm, but he liked them as they were; he didn't want to plant new things, nor to tear out old" (82). Maltby has a "long-visioned mind" (86); a free thinker, he does not feel the bind of any cherished preconceptions— nor does he feel the need to create any for himself. He tells his son, Robbie, that "in human thinking, bigness is an attribute of good and littleness of evil" (88). Maltby thinks in a most inductive manner. He and his lazy employee, Jakob Stutz, spend their days in idle contemplation. Robbie might make an observation, and the two men might use it "as the germ of an investigation" (87). "They were surprised at the strange fruit their conversation bore," Steinbeck writes, "for they didn't direct their thinking, nor trellis nor trim it the way so many people do" (87). The two men and the boy live a comfortable life, letting the land and themselves gently flow along the course of nature, finding a stray egg or weed-covered cucumber to eat. Maltby accepts what is, and although his farm has been overgrown by vines and overrun by varmints, he resides pleasantly in his house—never bothered by encroaching nature. Joseph Wayne may bleed to death trying to make some deep contact with the land, if not some kind of domination, but Maltby comes closer to it by a simple acquiescence to the natural.

Yet Maltby finally flees in defeat. His character leans too far to the side of passivity; the slightest pressure and he folds up. Steinbeck emphasizes his weakness from the beginning of the story; Maltby lives in San Francisco, "inextricably entangled in a clerkship, against which he feebly struggled for ten years" (80). His unhealthy lungs force him to leave the damp air of the city, and Maltby goes happily, thankful that some reason outside his own power has forced him to make a move he wanted but never could make on his own. His gentle laziness adds to the destruction of his wife and her two children; once she marries Maltby, he lets nature take its course—and it does so with a vengeance. Undernourished, the two boys quickly succumb to influenza and Mrs. Maltby—already weakened from delivering Maltby's child, Robbie— dies of the same disease soon after. He reads to the dying children, wandering vaguely from one to the other. However, against the black face of death, Maltby proves utterly inept. He does recognize his peculiar blindness: "There are long-visioned minds and short-visioned. I've never been able to see things that are close to me" (86). In short, he lacks pragmatism. Worse, he lacks courage, desire, and will.

Maltby's long view seems perfect for children but does not fare well in an adult world where only pragmatism will suffice. The beneficial side of his vision has a favorable influence on Robbie, whose powerful vocabulary and creative imagination quickly win over the schoolyard boys. Indeed, the ragamuffin style of the Maltbys takes the boys by storm—their innocent, open minds are ready to accept the freewheeling, new way of thinking Junius and Robbie offer. However, the adults—the respectable school board members, for example—cannot accept it. No one in the valley can tolerate laziness; the Maltby lifestyle runs counter to every preconception held by farmer and businessman, and Junius's lack of materialism marshals contempt. In a misguided attempt to do good, Mrs. Munroe buys clothes for Robbie and shames him: the boy realizes he is poor. Typical of his weakness, Maltby submits to the judgment of the townspeople, deciding that Robbie cannot exist free in nature, cannot run about without the proper attire, cannot—in the words of respectability—go on living in "squalor" (90). In the tragic final scene, Maltby and his son, uncomfortably donned in cheap new clothes, head off for San Francisco where the father will dutifully find a job as an accountant and do what the community would dictate is proper, probably a middle-class death sentence for the fragile Maltby.

Like Maltby, the paisanos of *Tortilla Flat* will not suffer from the overwhelming strength of their visions; there are neither Morgans nor Waynes in Steinbeck's 1935 novel. The paisanos have many of the fine qualities of Junius Maltby, wide-open thinkers (when they think at all) who abide by what is and who have an easy association with nature. They live with few preconceptions, ignoring the law and the church as necessary. Like the structure of the novel they inhabit, the paisanos live their days as a string of episodes. They exist as a group for a short time, finding their congregation at Danny's house a comfortable thing. Unlike Maltby, they have the added advantage of vigor—physically tough, no one could push around Danny, Pilon, Pablo, or Big Joe Portagee. When a small vision seizes them, they act aggressively but only on the short term, until the novelty wears off. They also have the strength of cooperation; "the friends" act as an ignoble band of knights, with Danny their King Arthur and an old house their Round Table. Only marginally selfish, each will give up some of his coveted wine to a friend, and when they sense real trouble—such as when Teresina's children run out of food—the paisanos demonstrate concern and compassion.

As paisanos, Danny and his friends "are clean of commercialism" (2). For all of their petty thievery, they lack the materialistic drive so fatal to other Steinbeck characters, from Henry Morgan to Ethan Hawley. Danny inherits two houses, and he reacts as Henry David Thoreau might; he simply feels "weighed down with the responsibility of ownership" (5). He wishes the houses were Pilon's and is relieved when one of them burns down, thus easing the burden of ownership from his shoulders. After Danny has died, the paisanos neither remain in the last house nor try to sell it off—they are content to watch it burn.

In their way, the paisanos are wise. Pilon, for example, is a man of observation and pragmatism; with his "fine clear vision" (32), he can rationalize any situation and justify his needs or the needs of his friends. When he sees an arm sticking out from under a bush with a jug of wine beside it, he quickly figures that the body attached to the arm, if dead, will not mind being relieved of the wine. When Pilon is faced with the problem of finding the Pirate's horde of quarters, he observes his subject's movements with scientific scrutiny and rationalizes any appropriation of the loot by deciding that he can offer the services of his brain to the brainless Pirate. Yet he has the paisano's heart of gold, for when the Pirate entrusts the money to Pilon and the others and explains that the money is meant to fulfill a promise to God, Pilon actually protects the quarters and discards his plan to spend them.

The paisanos can accept a truth that breaks George in *Of Mice and Men,* as well as any other Steinbeck character who succumbs to a strong vision: as Pablo says in chapter 14, "That is the way life goes, never the way you planned" (159). For the rest of the chapter, the paisanos relate tales of failed schemes, some laughable and some tragic. In many ways they live an ideal life, by nature's time, by "the great golden watch of the sun" (154).

All of these qualities would seem to point to the perfect group, men without preconceptions, delusions, or artificiality. Yet this band cannot last, nor can it make much impression on the world. Their way will not translate to others nor have any effect on the greater good, for it misses something important: progress. As we have seen, Ricketts and Steinbeck part ways over progress. However admirable "is" thinking may be, Steinbeck's desire to see humans take steps forward defies the passivity of mere acceptance. Breaking through to "is" brings one

closer to the truth, yet the best of Steinbeck's characters can do something with what they know. However admirable, the lifestyle at Danny's
house ultimately is a lotus-flower existence, a kind of stagnation. We
can agree with Astro, when he follows a line of criticism begun with
Peter Lisca, that the lifestyle of the paisano is a system in which nothing "will enable man to achieve and retain more than the most superficial of goals" (*Steinbeck and Ricketts* 112).

Typically, the paisanos do not mind the lack of progress—with the
exception of Danny. Noting the "changeless" quality of Monterey and
Tortilla Flat, the narrator observes that at "Danny's house there was
even less change" (169). Danny feels the monotony and knows he is
going nowhere: "Danny began to feel the beating of time" (170). He
rebels against the house and goes on a wild spree, doing all of his old
drinking, whoring, and thieving with desperate fury. When at last
spent, Danny is simply bored. "Do you know, Danny, how the wine of
your life is pouring into the fruit jars of the gods?," the narrator asks.
"Do you see the procession of your days in the oily water among the
piles?" (191). The narrator imagines the reflections of a cold historian
who might comment on Danny's insatiable lust at the last party: "A dying organism is often observed to be capable of extraordinary en-
durance and strength" (194).

Drunk, wild, and desperate, Danny terrifies the partiers with a rag-
ing desire to fight an "enemy who is worthy of Danny!" (196). He goes
out to battle the night air and dies after flinging himself off a forty-foot
cliff. With neither Danny nor the house to hold them together, the
paisanos separate, each to wander off alone. In the paisanos, Steinbeck
probes the opposite of the men of overwhelming vision—Danny and
his friends have everything but vision. Without it, and the great leaps
forward that vision might bring, there is no progress. Significantly,
Tortilla Flat concludes with a picture of stagnation and disintegration.
A life of perfect ease with the world can also be a trap; without struggle
brought on by the urge to push forward, there can be no evolution.

By 1938, Steinbeck comes close to his ideal with Billy Buck of the
stories in *The Red Pony* (first published in *The Long Valley* and then as
a separate book in 1945). Buck has strength, wisdom, purpose, and
great compassion. He lacks the greed of Morgan, the monomania of
Wayne, and the aimlessness of the paisanos. Buck is as essential a part
of the Tiflin ranch as he is of Jody Tiflin's life. Carl, Jody's father, has a
certain degree of wisdom, but his compassion is choked off by his

crushing pragmatism. Buck knows how a boy feels. However, Buck suffers from a flaw similar to Joseph Wayne's: he thinks he knows nature well enough to make it fit his plans and expectations. While the paisanos, in their easy acceptance of things, know the best-laid plans will go awry, Buck must endure some painful lessons in nature's school. To his credit, Buck does his best to learn, but his pride and his love for Jody get in the way. *The Red Pony*, like so much of Steinbeck's work, is an object lesson of our place in the world: no one dominates the natural scheme and no one can make promises against the greater machinations of the universe.

Buck is a strong, good ranch hand and has a degree of humility; he knows to wait on the steps rather than be the first in the dining room for breakfast, understanding his station as a hand. Yet Buck prides himself as an infallible teacher in the natural school; in the first pages of *The Red Pony* he instructs Jody on things from egg yolks to horse care. Jody, a wide-open boy eager to learn, accepts the ranch hand's word as perfect. Buck is in touch with nature, even explaining to Jody the need to converse with a horse to ease its fears (*Long Valley* 215). Such communion with nature indicates that, in Steinbeck's world, this is a good character—a man in tune with the whole.

Buck has set himself up in his own and Jody's eyes as an expert on nature: "Buck wasn't wrong about many things. He couldn't be" (222). That "couldn't be" suggests the trap Buck has set for himself, and the problems arising from the red pony, Gabilan, soon prove that even Buck cannot predict nature. He leaves the horse out in a storm, having erroneously assumed the weather would be fair. Although he predicts the horse will be all right, it becomes gravely ill and, despite all his efforts to save it, the red pony dies horribly. Jody strikes back at nature, killing one of the buzzards feasting on the dead pony—the act has little effect, however, for the bird's "red fearless eyes still looked at him, impersonal and unafraid and detached" (238). The boy and Buck know the gesture is useless, and they learn a lesson about promises, plans, and the greater processes of nature. The lesson does not hold.

Later in *The Red Pony*, in the chapter entitled "The Promise," Buck falls into the same trap as before. Trying to recover the position he has lost in Jody's eyes, he wants desperately to deliver Nellie's colt (which Tiflin has said will be Jody's). After warning that he cannot promise Jody anything, Buck soon feels proud again, telling Jody: "I'll see you get a good colt. I'll start you right" (*Long Valley* 273). To make his promise right, Buck has to kill and dismember Nellie to retrieve the colt from an impossible breach position. As Robert H. Woodward has shown, Buck becomes obsessed with trying to keep his promise to Jody. To fulfill the promise, nature has called for a bloody sacrifice, and neither the angered Buck nor the appalled Jody feel good about it.

Buck has made mistakes, but unlike Wayne, he is not destroyed by them. In the final chapter, "The Leader of the People," Jody eagerly plots the destruction of some mice, observing that they have no idea what will soon happen to them. "No, nor you either," Buck tells Jody, "nor me, nor anyone" (*Long Valley* 299). (In the original manuscript, Steinbeck crossed out a passage that equated the mice with people, personifying them as "mother mice, the caucuses of political mice, the gossiping cliques of social mice, the young fairy mice, all would go to the death hunted down by . . . Jody—Fate.") Just as Buck knows his place among men as a ranch hand, he must learn his place in nature as a human being. He has the wisdom and humility to accept the lesson and, unlike the paisanos, he can do something useful with it by passing it on to Jody.

THE "DOC" CHARACTERS

Aside from Tom Joad, to be discussed later, the evolution of the Steinbeck hero culminates in the various "Doc" characters. Whether Doc appears as Dr. Phillips of "The Snake," Dr. Burton of *In Dubious Battle*, or just plain Doc of *Cannery Row* and *Sweet Thursday*, Steinbeck handles him with respect and admiration—hardly surprising, since Doc is based on Ed Ricketts. Looking at the first three works Doc appears in, we can put together a character combining all of the best traits of Steinbeck's humanitarian. Doc suffers one important lack, however, and *Sweet Thursday* is especially interesting for its emphasis—strained to the point of nostalgia—on making Doc a whole man.

Doc embodies the good traits Steinbeck had been selecting out for a decade; the character is a humble, compassionate, strong, inductive seeker of truth and knowledge—a man in touch with the whole. Doc tries to see the wide picture and, therefore, cooperates with others and is quick to forgive. Above all, Doc is a patient teacher—his life as a gatherer of specimens for high schools and universities underscores his desire to observe and pass along his knowledge of the natural world.

The doctor also possesses physical strength; when provoked, he can fight with great fury.

In Doc's first appearance, as Dr. Phillips in "The Snake" (written in 1934 but included in 1938's *The Long Valley* [Benson, *True Adventures* 290]), Steinbeck shows how Doc's mind works. He strokes one of the cats in his lab before he drops it in a little gas chamber (the cat is earmarked for dissection) and we see his curious mix of compassion and rationality. When a strange, tall woman intrudes on his work, Doc forgets his irritation and includes her in what he is doing, trying to teach her about the sexual processes of starfish. Observing her cool attitude, he tries to upset her with his work on the dead cat—experimenting to see if this odd woman will react. When she asks to see a male rattlesnake, he tells her the difficulty of determining the snake's sex; casually, he notes a truth that others in Steinbeck's work have suffered for not knowing: "Nearly every generalization proves wrong" (75). The woman's sexual arousal from watching the snake consume a rat sickens Doc, for it crosses the line of his own sense of compassion: "He hated people who made sport of natural processes. . . . He could kill a thousand animals for knowledge, but not an insect for pleasure" (77). In this first sketch of Doc, as Dr. Phillips, we are introduced to his rational mind, his compassion, and his desire to learn and spread knowledge of the natural world. Something untrue, such as killing a rat for perverse pleasure, makes Doc ill.

Doc Burton of *In Dubious Battle* establishes the character's selflessness, further illuminates his inductive mind, and introduces Doc's one problem: loneliness. Upon Doc Burton's entrance into the novel, Mac sums him up: "He's a good guy. Looks like a pansy with his pretty face, but he's hard-boiled enough. And he's thorough as croton oil" (113). Doc risks his career as a doctor, and perhaps his life, by organizing sanitation at the strikers' camp. As Mac observes, "You're not a Party man, but you work with us all the time" (129). Doc demonstrates his penchant for taking action as needed, but in explaining to Mac his reasons for helping, he above all reveals the holistic quality of his thinking. He equates social injustice among men with physiological injustice among microbes. He comes to the strike to observe: "I want to see the whole picture—as nearly as I can. I don't want to put on the blinders of 'good' and 'bad,' and limit my vision" (130). Thus, Doc incorporates the same methodology of Steinbeck, Ricketts, and Darwin: go beyond self, beyond preconceptions, and see the whole. Doc sees the long view but also has an immediate perspective since he runs sanitation at the camp. And Doc can be assertive, for he has no fear of telling Mac what may be the truth, that the strike is "brutal and meaningless" (229).

Indeed, the process of seeing the whole, so effectively realized by Doc, also sets him apart from everyone else in the novel. Loneliness is the price of his total devotion to the truth, of trying to see *Homo sapiens* from the outside looking in. "I'm awfully lonely," Doc tells Mac, "I'm working all alone, towards nothing" (232). That Mac and the others cannot understand Doc's point of view shows how rare and alienating it can be. Foreshadowing *Sweet Thursday*, Mac observes that Doc "needs a woman bad" (233). However, judging by the manipulation, violence, and uselessness of the strike, Doc's perspective certainly seems superior. He has no one to share his thinking with, but the importance of his place in the strike is obvious once the Association picks him up; with Doc gone, the camp grows filthy, and the sheriff has an excuse, citing health code violations, to move in and scatter the strikers.

Steinbeck reintroduces Doc for 1945's *Cannery Row* and, in doing so, defines his conception of a hero: "Doc . . . is wiry and very strong and when passionate anger comes on him he can be very fierce. . . . [He] is half Christ and half satyr and his face tells the truth. . . . Doc has the hands of a brain surgeon and a cool warm mind. . . . Doc tips his hat to dogs as he drives by and the dogs look up and smile at him. He can kill anything for need but he could not even hurt a feeling for pleasure" (138). Being half Christ and half satyr, Doc holds the central vantage point, knowing and experiencing the best and worst of humanity. The "cool warm" mind suggests a rationality tempered by compassion. His amiable relationship with dogs indicates his general harmonious association with nature, born of his expanding knowledge of it.

All of these traits together allow him to know and relate the truth. The narrator describes how Doc first tried to tell people why he was taking a walking tour of the country: "Because he loved true things, he tried to explain" (214). His honest explanation, that he was walking to learn something about the country, was unacceptable to pragmatic listeners. (The story is based on Ed Ricketts's walking tour from Indiana to Georgia in 1920.) Always the teacher, Doc spreads his truth, like Burton of *In Dubious Battle*, to those who cannot understand it; the idiot, Hazel, baits Doc with questions just to keep conversation going. Doc knows this game but cannot help answering questions when they are asked (143).

When Mack and the boys try to throw a party for Doc (their one great vision and, typically Steinbeckian, a complete failure), Doc's patience is tested. They trash his lab, breaking hundreds of dollars worth of glass and equipment. Doc punches Mack, but his rage quickly fades, and he accepts the mess. Forgiving Mack, he holds out the truth that the boys will never pay him back, and shares a few philosophical beers with the crestfallen bum. Doc, like the narrator, knows the boys "are clean" of material desire and commercialism; he can forgive them because he respects their position in the big picture: "They just know the nature of things too well to be caught in that wanting" (252). For his wisdom, compassion, and forgiveness, Doc is an integral part of the Row, but he is still lonely and, in his detachment, has yet to savor that "hot taste of life" spoken of in "Black Marigolds," the poem he reads aloud at the end of the novel (306).

When he finished "Bear Flag" in 1953 (the book published as *Sweet Thursday* the next year), Steinbeck wrote, "It's crazy" (*Steinbeck: A Life* 473). The novel is indeed a crazy, nostalgic treatment of Doc and those human traits the author most cherished. The book reaffirms truth, holism, and the spirit of induction. Thematically, the craziness occurs because Doc, who has lost his way, is restored by a failed prostitute and a put-upon half-wit.

Many critics find the novel to be a slight work; recently, for example, Jackson J. Benson has ranked it with *Burning Bright* as examples of the novelist's worst writing (*Short Novels* 1). As Gladstein observes, "*Sweet Thursday* . . . is usually listed among Steinbeck's least-valued works" ("Straining for Profundity" 235). Richard Astro despairs somewhat at the incomplete Doc portrayed in *Sweet Thursday*, suggesting that Doc throws over his science—his seeking the "hot taste of life" in the Great Tide Pool—and succumbs to romance with Suzy. "The Doc of *Sweet Thursday* becomes a new man who no longer embraces Ricketts' way of thinking, and Steinbeck's depiction of his inability to achieve Ricketts' great emergent represents the novelist's dirge over the destiny of the Ricketts-like character in the modern world" (*Steinbeck and Ricketts* 199). Astro believes that Doc falls for "a self-satisfying type of romantic love" with the help of the Seer and—of all people—Joe Elegant (199). So poorly is the Doc of this novel received by critics that Astro writes, "Elaine Steinbeck has unequivocally denied that her husband set out to demean Ricketts in *Sweet Thursday*" (195).

Yet, underneath all of the craziness and fun, this last portrayal of Doc by Steinbeck is really the most loving, his highest tribute to Ricketts. The nostalgia of the novel derives from Steinbeck's desire to endow Doc with the widest sweep possible, to fix all of this character's problems. As such, *Sweet Thursday* is Steinbeck's most complete picture of the best-possible human being. The triumphant ride into the sunset that concludes *Sweet Thursday* is precisely the opposite of the bleak view of *In Dubious Battle*, where Doc is defeated and finally abducted by foes.

In many ways, Doc of *Sweet Thursday* is an extension of the character in previous works; he is a great sympathizer, cooperator, thinker, and teacher. Steinbeck chooses to emphasize Doc's honesty in this novel, which is not surprising since from *The Wayward Bus* to *The Winter of Our Discontent* the author has portrayed his disgust with what he perceived as the lack of truth in American society. But from the scientific perspective, from which Steinbeck never strays too far, honesty is a cornerstone of the objective search for truth. On several occasions, Doc worries about dishonesty, raising questions of legality and integrity to Wide Ida, Fauna, and Old Jingleballicks (84, 133, 171). Joseph and Mary Rivas, whom Doc eventually beats in a fight, exists as a foil to the honest man: "Doc and Joseph and Mary were about as opposite as you can get. . . . Doc was a man whose whole direction and impulse was legal and legitimate" (13). Rivas is confused and fascinated by Doc's honesty, just as he is amazed by Doc's game, chess, which relies on strength of mind and cannot be rigged.

But the loneliness that has been Doc's bane since *In Dubious Battle* erupts into a crisis in *Sweet Thursday*. (Ricketts rarely suffered from loneliness, as he was married once and lived with another woman for seven years; however, his relationships with women could go terribly sour, as one of his last letters to Steinbeck, written in November of 1947, describes at length). The fictionalized Doc has a song of loneliness welling up within him, one that he can no longer ignore. Writing a paper on octopi or throwing himself into his work cannot stifle the growing need to be with someone. Although Astro believes that for Doc the "hot taste of life" exists in the fecundity of the Great Tide Pool (203), this observation is only partially true. A man as distant from his own species as Doc has been must finally suffer loneliness, for life as an observer of his or other species ultimately has limited satisfaction. Doc senses that for all of the wide view science has given him, the aperture is still not wide enough, and it is more than a mere joke when the boys ignorantly buy Doc a telescope instead of a microscope. Doc needs new directions; the great cooperator, the great humanitarian, should have someone with whom to share his life. His "top mind," the director of his scientific methodology, becomes a tyrant: "No single thing could be permitted in unless it could be measured or tasted or heard or seen. The laws of science were Doc's laws, and he sought to obey them" (235). Even the best method, if it becomes too rigid, can grow into its own trap—a set of rules, another preconception, a warp. Doc cannot solve his own problem of loneliness; everyone from Mack to Fauna knows he needs a woman, but how might a relationship be arranged?

First, the right woman must appear. Steinbeck clearly intends Suzy to be an ideal match for Doc. She too is honest and even less likely to compromise her honesty than Doc: she tells him that he will never write his paper. Feeling bad for him, she tries to take it back, but Doc agrees, "I love true things . . . Even when they hurt. Isn't it better to know the truth about oneself?" (108). "She slapped me in the face with a few basic truths," Doc later tells Fauna (131). Suzy is a realist, observant, inquisitive, and tough—in fact, after Cathy Ames, she is Steinbeck's most independent female character. She sets up her house in the boiler, and when Doc sees how she lives, he thinks, "My God, what a brave thing is the human!" (247). Suzy embodies all the elements of Steinbeck's best people, including humility, for she does not think herself good enough for Doc, and compassion, for his broken arm brings her immediately to his side despite all previous barriers.

Still, the problem of getting the two together remains. *Cannery Row* invests in mystic and romantic solutions, which—considering the body of Steinbeck's work—are doomed to be inadequate. Fauna looks to astrology for answers, a science Suzy feels is merely "that crap about stars" (114). Fauna does her best, although her information is faulty; she believes Doc to be a Cancer, born in July, but the narrator reveals later that he was born in December. The stars lead Fauna to organize a romantic costume party in which Suzy, uncomfortable in her virginal Snow White dress, meets an equally ill-at-ease Doc. The party fails to bring them together, and the hopes of the Row are dashed—yet another vision, constructed of delusive mysticism and romanticized planning, that ends in disaster. No one on the Row knows what to do, and Doc is too emotionally involved to see out of himself and solve the problem rationally.

So, in one of Steinbeck's most outrageous and hilarious twists, the idiotic Hazel solves the problem and finds greatness by using the inductive method! As we know from his introduction in *Cannery Row*, Hazel's mind is entirely wide open—too wide, in fact. He suffers no preconceptions or narrowing of his views because he has not bothered to form any. Hazel also lacks an ego. He does spend much of his time listening, although people bare their souls to him only because they know he will not record what he hears and so will never pass on their secrets. And Hazel possesses some compassion: he loves Doc.

Once Fauna's astrological prediction reveals that Hazel will one day be president, the poor man must bear the weight of this new responsibility and concludes that he will have to change his way of thinking if he is to handle affairs in Washington. Hazel's first try, when he attempts to learn what makes Doc feel bad, is a great failure because of his faulty deductive method. He deduces that if Doc is in trouble, some person must be responsible and, since Mack has spread the word that Doc cannot write a paper, Hazel decides Mack is to blame. The answer: hurt Mack, perhaps even murder him. Hazel operates on a false theory based on shaky deduction, but fortunately Mack and the boys talk him out of his dangerous solution.

Confronted with the problem of bringing Doc and Suzy together, Hazel then adopts an inductive method, and predictably, considering Steinbeck's perspective, the idiot succeeds where everyone else has failed. In the chapters "A President Is Born" and "The Thorny Path of Greatness," Hazel hits upon his inductive process. Feeling the weight of his impending presidency, Hazel is "set apart and above ordinary experience" (212). Like Doc, this isolated position makes him feel lonely but also lets him see the truth from the outside. He decides to ask questions and this time learn from them: "He would not only listen, he would remember, and he would put all the answers together" (217). He goes out collecting data. Joe Elegant, the effeminate symbolist whom the narrator mercilessly ridicules, buries his answer to Doc's problem in a ludicrous series of symbols and concludes, wrongly, that Suzy "would only be a new false path" for Doc (220). Yet, by telling Hazel that Doc "needs love," Elegant provides a useful specimen for examination (220). An interview with Fauna reveals that if Suzy would go collecting at La Jolla with Doc, things would work out. Rivas, selfishly concerned with his own problems, only reveals a lusty personal interest in Suzy. Talking with Suzy herself, Hazel learns that she would only come to Doc if he "was sick or bust his leg" (228). Hazel visits Doc, who tells him, "It might turn out you know about people too"—a prophetic assessment (230). Hazel learns how Doc likes Suzy, and the idiot cleverly tells Doc that Joseph and Mary is "hanging around the boiler" (232). This information lets Doc's passion loose and leads to the fight outside Suzy's door. Hazel finally sees Mack, who wonders what the fool has been doing. "I bet right now you're trying to figure out germ warfare," Mack tells Hazel, reacting to what Hazel has reported about Suzy's belief that she could only see Doc if he were hurt (234).

With all of the necessary information, Hazel goes to the Seer (in jail for stealing candy bars) and tests his solution by posing the Seer a "hypothetical question" (254). The Seer confirms the hypothesis: "If you love him you must do anything to help him—anything. Even kill him to save him incurable pain" (254). Having gathered the data and tested the hypothesis, Hazel executes the solution: he breaks Doc's arm with an indoor baseball bat. Suzy rushes to Doc's aid, they declare their love for one another, and Hazel's theory proves correct. Knowing what Hazel has done, Mack watches Suzy and Doc ride off and, putting his arm around Hazel, says in the last line of the novel, "I think you'd of made a hell of a president" (272).

Hazel's momentary greatness, attributable to his use of the inductive method, confirms Steinbeck's belief in this mode of thought and also in the inductive "flame of conception," which earlier in the novel he credits Darwin, among other greats, with possessing. Considering that it was Hazel who found the answer, Steinbeck's comments on "the inductive leap" seem appropriate: "Everything falls into place, irrelevancies relate, dissonance becomes harmony, and nonsense wears a crown of meaning" (28). The answer brings Doc and Suzy together, two special people who will form a greater whole; indeed, Doc exclaims that he needs her in order for him to be whole (243).

Doc leaves his life on the page a fully realized hero—a man embodying the entire range of human potential, from passion to science. That he and Suzy are headed for the tide pools at La Jolla and a greater research facility provided by Old Jingleballicks means that Doc will not abandon his scientific life—the way is paved for greater discoveries by a mind opened ever wider. In the poignant last scene, Doc turns the corner to begin a mission of discovery, instead of meeting death on the railroad tracks as Ricketts did. *Sweet Thursday* by no means denigrates

Ed Ricketts; rather, it is the author's wish book for his lost friend, his real-life hero.

❖ ❖ ❖

From Henry Morgan to Doc, Steinbeck tested the nature of his protagonists and found his brighter vision of humanity. By the time he wrote *East of Eden,* his conception of the hero had been fully developed; the wide-open, compassionate character of Samuel Hamilton reads as if the narrator were describing an old friend (Steinbeck did base the character on his maternal grandfather). The author's vision of humanity's future lies in such characters as Doc, Hamilton, and Tom Joad: standing above the beast of *Homo sapiens,* a humble observer who thinks inductively, who searches for truth to pass it on, who cooperates with others, whose sympathy reflects a wider view, and whose self transcends prejudice and selfish borders. Such characters have connections to the earth, know their place upon it, and draw strength from it. In many ways, Steinbeck's hero parallels the youthful Charles Darwin that the author knew by reading the *Beagle* journal. No wonder Steinbeck reveres Darwin in *Sea of Cortez,* for that book came shortly after *The Grapes of Wrath,* the novelist's most Darwinian work.

The Darwinian *Grapes of Wrath*

"I often bless all novelists."
—*Charles Darwin,* The Autobiography of Charles Darwin, *1892*

A study of Charles Darwin and the art of John Steinbeck must, like any expedition through the novelist's life work, finally arrive at his master-piece, *The Grapes of Wrath.* In no other book is Steinbeck's dramatiza-tion of Darwin's theory more clear; the novel resonates with the natu-ralist's ideas. Through Steinbeck's narrative technique, from the parts (i.e., the characters in the Joad chapters) to the whole (the intercalary chapters), we are presented with a holistic view of the migrant worker developed through Steinbeck's own inductive method. This epic novel demonstrates the range of Darwin's theory, including the essential as-pects of evolution: the struggle for existence and the process of natural selection. The migrant workers move across the land as a species, up-rooted from one niche and forced to gain a foothold in another. Their struggle is intensified by capitalism's perversion of natural competition, but this only makes the survivors that much tougher. Because of their inability to see the whole picture, the bankers and members of the Farmers Association diminish themselves by their oppressive tactics while the surviving migrant workers become increasingly tougher, more resourceful, and more sympathetic. Ultimately, seeing Darwin's ideas in *The Grapes of Wrath* enables us to perceive some hope for the Joads and others like them—here is Steinbeck's manifesto of progress, based on biological laws rather than political ideology. Despite the dis-mal scene that concludes the book, we come to a better understanding of what Ma Joad already knows, that "the people" will keep on coming.

Steinbeck embarked on an expedition of his own from 1934 to 1938 to gather information that would ultimately lead to his great novel. Jackson J. Benson's biography of Steinbeck provides a very complete and accurate account of the novelist's research and, in direct reference to *The Grapes of Wrath* itself, Robert DeMott's introduction and notes for *Working Days* provide further illumination and detail.

Benson writes that Steinbeck, who "seems to have remembered in detail nearly everything he saw or heard," entered the world of migrant labor in California when he interviewed two starving, fugitive strike organizers in Seaside in early 1934 (*True Adventures* 291). Steinbeck gathered more information from James Harkins, an organizer who helped in the Imperial Valley strike (1934) and the Salinas lettuce strike (1936). Eventually strike organizers began to frequent the Steinbecks' cottage in Pacific Grove and discuss their "holy mission": "Since you were either for them or against them—there was no compromise—[Steinbeck] did more listening than talking" (294). Steinbeck also met the famous social reformer and muckraker, Lincoln Steffens, who was spending his last years in a house in nearby Carmel. Benson writes that Steffens and Steinbeck agreed on "the importance and value of observing and discovering" (295). Steinbeck, as early as the summer of 1934, had himself gone out to see the migrant labor camps in the Salinas area. All of the information he gathered, along with very detailed information from a union leader, Cicil McKiddy, eventually became a part of *In Dubious Battle*. Significantly, the book does not follow anyone's party line but rather works out many of Steinbeck's and Ed Ricketts's biological views.

Serious research for *The Grapes of Wrath* began with Steinbeck's assignment for *The San Francisco News* to write a series of articles about migrant farm labor in California, which necessitated observing conditions at various labor camps. He saw firsthand the destitution of migrant families in these government camps and spontaneous Hoovervilles. As Benson and DeMott show, a tremendous influence on Steinbeck as he prepared to write *The Grapes of Wrath* was Tom Collins, the manager of "Weedpatch," the government Sanitary Camp at Arvin (he is the "Tom" that the book is partly dedicated to). Collins was something of a social scientist who made meticulous reports and gathered statistics about the migrant's life which Steinbeck used extensively in *The Grapes of Wrath* (Benson, *True Adventures* 343-44). Even at home near Salinas, Steinbeck found more information to gather, as incidents of vigilantism were occurring as a result of the strike of 1936 (346). In 1937 Steinbeck took another, longer tour of migrant camps with Collins, and in February of 1938, Steinbeck went to the flooded areas of Visalia where, as he wrote to his agent, "Four thousand families, drowned out of their tents are really starving to death" (368). As DeMott writes, "What he witnessed there became the backdrop for the final scenes of *The Grapes of Wrath*" (*Working Days* 134).

When the author began to write up his observations into a novel, his first bitterly satirical attempt, "L'Affaire Lettuceberg," failed because he was too close to the subject. Like Darwin's *Origin, The Grapes of Wrath* is a gathering of observations fused by a hypothesis, in this case a biological consideration of cycles in land ownership. Of course, unlike *Origin*, it is fictionalized and, above all else, a work of art. Still, Steinbeck's method in putting together the novel resembles an inductive, scientific one. Anything less, in the hands of some other writer, might have been another political satire like the "L'Affaire Lettuceberg."

From the first pages of *The Grapes of Wrath*, Steinbeck's biological, holistic view is evident. The novel presents a large picture in which humans are only a small part; in the great natural scheme of sky and land, of rain, wind, and dust, they suffer with the teams of horses and the dying corn—all life forms are helpless in this huge canvas of natural machinations. And there are beasts at the door; not more than a night after the people leave Oklahoma enter new occupants who were always waiting outside: weasels, cats, bats, and mice (*Grapes* 126-27).

People are further associated with the natural world by being rendered in animal metaphors, either by their own language or the narrator's. In chapter 8, we meet the Joad family and hear that Ma fears Tom will be like Pretty Boy Floyd ("They shot at him like a varmint . . . an' then they run him like a coyote, an' him a-snappin' an' a-snarlin', mean as a lobo"); that Grampa once tortured Granma "as children torture bugs"; that Grampa had hoped the "jailbird" Tom would "come a-bustin' outa that jail like a bull through a corral fence"; and that somewhere Al is "a-billygoatin' aroun' the country. Tom cattin' hisself to death" (82-89).

The narrator's famous image of the land turtle is the most extensive metaphor for the migrant worker. In chapter 3, the tough, wizened turtle navigates the road, pushing ahead with "hands" rather than front claws. Tom picks up the turtle and Casy observes, "Nobody can't keep a turtle though . . . at last one day they get out and away they go—off somewheres. It's like me." (21). When Tom releases it, a cat attacks it to no avail, and the turtle goes in the same direction that the Joads will: southwest. The connection is made even stronger when, in chapter 16, a description of the flight of the Joads and the Wilsons across the

Panhandle is juxtaposed with the image of the land turtles which "crawled through the dust" (178). Steinbeck's extensive use of personification and anthropomorphism underscores his view of *Homo sapiens* as just another species.

This recognition leads to the same collision with traditional religion that Darwin's theory encountered in Victorian England, by directly challenging the idea that the human is above the animals, a being made in God's own image. In an eerie scene, Steinbeck powerfully demonstrates the self-delusion of a group of "Jehovites" who pray in a tent for Granma. Aspiring to be superior to the natural world, they are more beastlike than those they call sinners: "One woman's voice went up and up in a wailing cry, wild and fierce, like the cry of a beast; and a deeper woman's voice rose up beside it, a baying voice, and a man's voice traveled up the scale in the howl of a wolf. The exhortation stopped, and only the feral howling came from the tent" (233). Like meetings Casy devised as a preacher, in which excited men and women went from the meeting place to the bushes to make love, traditional religion is only another veneer over animal nature.

Certainly the world Steinbeck portrays in *The Grapes of Wrath* demonstrates what Darwin, Ricketts, and he believed: humans are subject to the laws of ecology. That the Darwinian principles of competition and selection are an essential part of the novel is no surprise. The Joads and Wilsons are part of a movement of migrants, acting as a species turned out of a niche by natural and unnatural forces. The migrants go to a richer niche that would appear to have plenty of room for them, but many of them die, overwhelmed by competition and repression. Yet the survivors display an astounding ability to adapt. They come to California, a vigorous new species quite terrifying to the natives who, despite the crushing power of a brutal economic system which they control, act from a growing sense of insecurity. "They have weathered the thing," Steinbeck writes of the migrant workers, "and they can weather much more for their blood is strong. . . . [T]his new race is here to stay and . . . heed must be taken of it" (*Gypsies* 22).

The process of evolution that leads to the creation of "this new race" is patently Darwinian. With drought upon the land and the dissolution of the tenant system, the farmer can no longer live in the region—forcing the migration west. In *The Origin of Species*, Darwin observes that if an open country undergoes some great change, "new forms would certainly immigrate, and this would likewise seriously disturb the relations of some of the former inhabitants" (Appleman 55; see also 97). From the first day of the Joads' migration, a process of selection begins; those who can adapt to the new way of life survive. Although a tough man, Grampa proves too rooted in the old land to adapt to the new, and his death, as Casy knows, is inevitable: "Grampa didn' die tonight. He died the minute you took 'im off the place" (*Grapes* 160). Muley cannot leave either, and his future is doubtful; ironically, Noah, who himself will wander off alone into oblivion, tells Muley, "You gonna die out in the fiel' some day" (121). Granma cannot recover from the death of Grampa and loses touch with reality and eventually life. The Wilsons also fail, despite help from the Joads; Ivy lacks the essential mechanical knowledge of cars to succeed, and Sairy is too physically weak to survive.

Because of the migrants' relentless trek, during which they are driven by the harshness of the weather, by poverty, and by cruelty, the ones who arrive in California already are transformed. As intercalary chapter 17 shows, the group has adapted to the new way of life on the road: "They were not farm men anymore, but migrant men" (215). The new breed pours into California "restless as ants, scurrying to find work to do." But "the owners hated them because the owners had heard from their grandfathers how easy it is to steal land from a soft man if you are fierce and hungry and armed" (256-57). Ma Joad typifies the strong blood that Steinbeck refers to, for she adapts to each new situation, meeting difficulties with whatever ferocity or compassion is needed, constantly working to keep the family together and push them forward. Toward the end of the novel, Ma gives her famous speech about the people, and certainly she has come to understand what survival of the fittest means: "We ain't gonna die out. People is goin' on—changin' a little, maybe, but goin' right on . . . some die, but the rest is tougher" (467-68).

A Darwinian interpretation of *The Grapes of Wrath* reveals the novel's most terrible irony: the owners' perversion of the natural process only hastens their own destruction. In states such as Oklahoma, the bank—the "monster"—must be fed at the expense of the tenant system, thus losing something precious: "The man who is more than his chemistry . . . that man who is more than his elements knows the land that is more than its analysis" (126). And the tenacity of these people, their potential, is drawn to another land. The Farmers Association of California sends out handbills to attract a surplus of labor, intensifying the competition for jobs so that the migrant laborers will work for almost nothing. But the owners are unconscious of the other part of the equation, that increased competition only toughens the survivors, as Darwin notes: "In the survival of favoured individuals and races, during the constantly recurrent Struggle for Existence, we see a powerful and ever-acting form of Selection" (Appleman, *Origin* 115). The novel's omniscient narrator recalls "the little screaming fact" evident throughout history, of which the owners remain ignorant: "repression works only to strengthen and knit the repressed" (*Grapes* 262).

Steinbeck recognizes the untenable position of the owners in California. "Having built the repressive attitude toward the labor they need to survive, the directors were terrified of the things they have created" (*Gypsies* 36). As the economic system blindly pushes people out of the plains states and just as blindly entices them to California with the intention of inhumane exploitation, it proves a system of men who fail to see the whole. Often the owners win, and some workers are hungry enough to betray their own kind, such as the migrants hired to move in and break up the dance at Weedpatch. But at the end of chapter 19, the omniscient voice describes how in their suffering people come together, as the migrants gather coins to bury a dead infant; soon they will see beyond themselves and the illusion of their religion: "And the association of owners knew that some day the praying would stop. And there's the end" (*Grapes* 263).

The narrator describes the sense of coming change in more ominous tones at the end of chapter 25: "[I]n the eyes of the hungry there is a growing wrath. In the souls of the people the grapes of wrath are filling and growing heavy, growing heavy for the vintage" (385). The narrator presents the whole view, which characters like Casy and Tom eventually see but the owners remain blind to as they continue to create a breed that will be their undoing. "For while California has been successful in its use of migrant labor," Steinbeck writes, "it is gradually building a human structure which will certainly change the State, and may, if handled with the inhumanity and stupidity that have characterized the past, destroy the present system of agricultural economics" (*Gypsies* 25). This is the dynamic that Steinbeck describes in *The Log* after visiting Espiritu Santo Island, where in certain areas only one or two species dominate an ecosystem. He parallels the territorial habits of these animals with humans. While the "dominant human" grows weak from too much security, "[t]he lean and hungry grow strong. . . .

Having nothing to lose and all to gain, these selected hungry and rapacious ones develop attack rather than defense techniques . . . so that one day the dominant man is eliminated and the strong and hungry wanderer takes his place" (97).

In the thinking of Darwin and Steinbeck, the California landowners' "inhumanity" is their keen lack of sympathy and their "stupidity" is the reason for that lack, the inability to see the whole. As Casy tells his assassin just before the death blow, "You don't know what you're a-doin" (*Grapes* 426). His last words appropriately echo Christ's, for in killing the leader of a cause, one leaves tougher disciples, such as Tom Joad.

From his knowledge of the whole, of past and present, and of humanity's true place in the scheme of nature, Charles Darwin nears the end of *The Descent of Man* with the interesting realization that he would rather be a "heroic little monkey" than the human "savage who delights to torture his enemies . . . knows no decency, and is haunted by the grossest superstitions" (Appleman 208). It is not the kind of statement that anyone with illusions about the inherent superiority of human beings would wish to hear. Darwin's view is certainly played out in *The Grapes of Wrath,* as we encounter a group of the most civilized people practicing many of the atrocities that delight Darwin's savage. They lack sympathy and therefore will lose their humanity and probably their existence as a group. In contrast, the migrant workers show a sense of compassion for their fellows that binds them together and can eventually insure their existence in a hostile environment, for the cooperation that grows out of sympathy is the greatest threat to the owner, as Steinbeck forsees: "And from this first 'we' there grows a still more dangerous thing: 'I have a little food' plus 'I have none.' If from this problem the sum is 'We have a little food,' the thing is on its way, the movement has direction. . . . If you who own the things people must have could understand this, you might preserve yourself" (*Grapes* 165-66). Steinbeck goes on to warn that "the quality of owning freezes you forever into 'I,' and cuts you off forever from the 'we'" (166). Clearly, the owners do not understand this reality.

The other great irony of the novel is that, through a Darwinian process of adaptation and evolution, the dehumanizing conditions created by the owners only make the migrant workers more human. This process can be seen in nearly every chapter, as migrants share money, food, transportation, work, and ultimately their anger, as they briefly unite in a strike that is defeated by an influx of hungry workers who do not yet see the big picture. But as the suffering continues and more Casys are martyred and more Toms are created, the people will eventually move forward. As Casy tells Tom, "ever' time they's a little step fo'ward, she may slip back a little, but she never slips clear back" (425). Casy's words resonate with the narrator's definition of what man is in chapter 14: "This you may say of man . . . man stumbles forward, painfully, mistakenly sometimes. Having stepped forward, he may slip back, but only a half step, never the full step back" (164). While the owners, comfortable and rich, are frozen in their "I" mentality, the surviving migrants move forward; they are vigorous and continue to evolve into their "we" mentality. This process is the essence of Steinbeck's scientific, Darwinian belief in a progression for humankind based on biological principles generally and struggle in particular. This is why, by the time he wrote *America and Americans* in 1966, he worries most of all that the country has lost its survival drive.

This particular kind of evolution is best illustrated through the development of Tom Joad, perhaps Steinbeck's most complete hero. Although Ma, too, shows a tremendous capacity for adaptation and sympathy, her sense of "we" does not extend much beyond the family unit, and while Casy certainly comes to see the whole picture and extends his sympathy to all oppressed laborers, the greatest change occurs in Tom, whose near-animal introversion becomes an almost spiritual extroversion during his family's struggle to survive. This change in character has been noted by several critics. Lisca calls Tom's conversion one from the personal/material to the ethical/spiritual ("Grapes" 98). Charles Shively believes that Tom's widening horizon reflects the influence on Steinbeck of American holistic philosopher Josiah Royce (there is no evidence the novelist had heard of Royce before 1948, however [see DeMott's *Steinbeck's Reading* 96, 169]). And Leonard Lutwack sees Tom's conversion through Biblical imagery, from his "baptism" when he kills Casy's assailant by a stream to his "resurrection from the tomb" while he speaks to his mother in the cave (70-71).

Tom has received so much attention from critics probably because he is Steinbeck's most dynamic character. At the beginning of the novel, Tom—like Grampa, Al, Ruthie, and Winfield—is preoccupied with his own needs. Sitting with Casy and Muley in his parents' wrecked house, Tom has only food on his mind while Casy talks on about their pathetic existence: "Joad turned the meat, and his eyes were inward" (*Grapes* 54). Once the meat is done, Tom seizes it, "scowling like an animal" (57). Casy suddenly is inspired to go with the people on the road, but Joad merely rolls a cigarette and ignores the preacher's speech. Later, Tom rails at a gas station attendant who worries about what is happening to the country, then pauses, noticing for the first time that the attendant's station is near bankruptcy. Tom corrects himself, "I didn't mean to sound off at ya, mister" (139). When Casy questions Tom about the larger picture, about the fact that their group is only a small part of a mass migration, Tom begins to feel the inadequacy of his narrow vision: "I'm jus' puttin' one foot in front a the other. I done it at Mac for four years. . . . I thought it'd be somepin different when I come out! Couldn't think a nothin' in there, else you go stir happy, an' now can't think a nothin'" (190). Like Ma, keeping the family together becomes a project for Tom; he finds his drunken Uncle John and feels pity for him, and later he sacrifices his natural anger for the good of the family (306, 309).

After Tom sees Casy killed, the change that has been gradually occurring in him becomes complete. He understands the entire structure, for he has been both in the camp with the laborers and outside with the strike leaders. His immediate concern is to flee to protect his family, but to appease Ma he hides in a cave of vines near the Joads' new camp. After Ruthie spills the news that Tom killed Casy's assassin, Ma goes to him in the cave for a final talk. Although he has been reduced to living like an animal, in the darkness of his cave Tom has been thinking about Casy: "He talked a lot. Used ta bother me. But now I been thinkin' what he said, an I can remember—all of it" (462). He has come to realize the truth of the "Preacher," a code of survival based on cooperation: "And if one prevail against him, two shall withstand him, and a three-fold cord is not quickly broken" (462). He gives a speech in which he becomes the ultimate expression of sympathy, for he is now less an individual and more the essence of the whole: "I'll be ever'where—wherever you look" (463). He has determined a truth that goes beyond even Ma's comprehension, for she says at the end of his speech, "I don' un'erstan" (463). Because of his fierceness, inherited from Ma, he will pose a greater threat to the owners than Casy, who "didn' duck quick enough" (463).

In a work so full of apparently hopeless suffering, the Darwinian view of *The Grapes of Wrath* explains why characters such as Ma or Tom have a sense of victory. The processes of competition and natural selection, artificially heightened by narrow-minded landowners, create a new race with strong blood—a race that can adapt and fight in a way the old one could not. Endowed with a closeness to the land and an increasing sympathy, this new race represents a human being far superior to the old "I" savage. Because of the struggle, people like the Joads become better human beings, cooperating with each other in every crisis. "There is a gradual improvement in the treatment of man by man," Steinbeck wrote in a letter for the *Monthly Record* (a magazine for the Connecticut state prison system) during the period in which *The Grapes of Wrath* was being written. "There are little spots of kindness that burn up like fire and light the whole thing up. But I guess the reason they are so bright is that there are so few of them. However, the ones that do burn up seem to push us ahead a little" (3). Thus, even when famished and facing death herself, Rose of Sharon begins to see past her own selfishness and offers her breast to a starving man. She has reason to smile mysteriously, understanding something larger and greater than her oppressors will ever know.

❖ ❖ ❖

On the morning of May 15, 1992, Professor Stanley Brodwin of Hofstra University stood in the historic Admiral Coffin School hall on Nantucket Island and gave a lecture titled "The Example of Darwin's *Voyage of the Beagle* in *The Log from the Sea of Cortez.*" His talk was that rare recognition of Darwin's influence upon a book by Steinbeck. Brodwin discussed *The Log* as part of the larger genre of journals of expeditions by such naturalists as Darwin, Alexander von Humboldt, and Edward Forbes. He examined Steinbeck's book on another level among the four Steinbeck professed to be in it: "*The Log* remains a fully romantic work, its theological explorations maintaining a meaningful tension with its search for hard scientific information."

When Brodwin finished, the old hall was filled with applause. The audience's enthusiasm was not surprising, for he read at a conference, "Steinbeck and the Environment," co-sponsored by the University of Massachusetts Nantucket Field Station and the John Steinbeck Research Center. With Elaine Steinbeck seated in the front row, the rest of the hall was filled with literary critics, biologists, and environmentalists from England, Japan, and the U.S. When the sessions were over, and papers such as Richard Astro's "Discovering Ricketts: Marine Science Enters the English Department" and Warren French's "How

Green was John Steinbeck?" had been read, the literary scholars and scientists ambled up the cobbled roads into Nantucket Village with the author's wife near the head of the group.

It was a good day for Steinbeck studies, devoted to the truly deep element in his many-layered vision of humanity. His biological perspective came up again and again. When the conference moved to the Nantucket Marine Laboratory, where participants could examine experiments conducted by Dr. Joseph Grochowski and his associates, they peered into the bubbling tanks of marine life or looked out toward sun-dappled ripples in the harbor and talked of marine biology and the novelist's art. Had he been there, John Steinbeck would have been pleased.

BIBLIOGRAPHY

Primary Sources

Darwin, Charles. *The Autobiography of Charles Darwin and Selected Letters.* Ed. Francis Darwin. 1892. New York: Dover Publications, 1958.

———. *The Origin of Species* and *The Descent of Man. Darwin.* Ed. Phillip Appleman. New York: Norton, 1979.

———. *The Origin of Species* and *The Descent of Man.* New York: The Modern Library, n.d.

———. *The Structure and Distribution of Coral Reefs.* 1842. Berkeley: U of California P, 1962.

———. *The Voyage of The Beagle.* New York: Penguin, 1988.

———. *Voyage of The Beagle.* Ed. Janet Browne and Michael Neve. New York: Viking Penguin, 1989.

Steinbeck, John. "About Ed Ricketts." *The Log from the Sea of Cortez.* 1951. New York: Penguin, 1975. vii-lxiv.

———. *America and Americans.* New York: Viking, 1966.

———. *Bombs Away.* New York: Viking, 1942.

———. *Burning Bright.* 1950. New York: Penguin, 1979.

———. "The Chrysanthemums." *The Long Valley.* 1938. New York: Penguin, 1986. 1-18.

———. *Cup of Gold.* 1929. New York: Penguin, 1976.

———. *East of Eden.* New York: Viking, 1952.

———. Foreword. *Between Pacific Tides.* By Edward F. Ricketts and Jack Calvin. 1939. 3rd ed., rev. Ed. Joel W. Hedgpeth. Stanford: Stanford U P, 1952. v-vi.

———. *The Grapes of Wrath.* 1939. New York: Penguin, 1976.

———. *The Harvest Gypsies.* 1936. Berkeley: Heyday Books, 1988.

———. *In Dubious Battle.* 1936. New York: Penguin, 1979.

———. "Johnny Bear." *The Long Valley.* 1938. New York: Penguin, 1986. 141-66.

———. *Journal of a Novel.* New York: Viking, 1969.

———. "The Leader of the People." *The Long Valley.* 1938. New York: Penguin, 1986. 281-304.

———. Letter. *Monthly Record.* June 1938: 3.

———. *The Log from the Sea of Cortez.* 1951. New York: Penguin, 1975.

———. *The Long Valley.* 1938. New York: Penguin, 1986.

———. *The Moon Is Down.* 1942. New York: Penguin, 1987.

———. *Of Mice and Men/Cannery Row.* 1937, 1945. New York: Penguin, 1987.

———. "The Murder." *The Long Valley.* 1938. New York: Penguin, 1986. 167-84.

————. *Once There Was a War.* 1958. New York: Penguin, 1977.

————. *The Pastures of Heaven.* 1932. New York: Penguin, 1986.

————. *The Pearl.* 1947. New York: Penguin, 1986.

————. *The Pearl/The Red Pony.* 1947, 1945. New York: Penguin, 1986.

————. "A Postscript From Steinbeck." *Steinbeck and His Critics: A Record of Twenty-five Years.* Ed. E.W. Tedlock and C.V. Wicker. 1957. Albuquerque: U of New Mexico P, 1969. 307-08.

————. "The Raid." *The Long Valley.* 1938. New York: Penguin, 1986. 89-105.

————. *The Red Pony. The Long Valley.* 1938. New York: Penguin, 1986. 201-304.

————. *A Russian Journal.* New York: Viking, 1948.

————. *The Short Reign of Pippin IV.* 1957. New York: Penguin, 1977.

————. *Selected Essays of John Steinbeck.* Ed. K. Nakayama and H. Hirose. N.p.: Shinozaki Shorin P, 1981.

————. "The Snake." *The Long Valley.* 1938. New York: Penguin, 1986. 67-82.

————. *Steinbeck: A Life in Letters.* Ed. Elaine Steinbeck and Robert Wallsten. 1975. New York: Penguin, 1989.

————. *Sweet Thursday.* 1954. New York: Penguin, 1986.

————. *To a God Unknown.* 1933. New York: Penguin, 1987.

————. *Tortilla Flat.* 1935. New York: Penguin, 1986.

————. *Travels with Charley in Search of America.* 1962. New York: Penguin, 1980.

————. *Uncollected Stories of John Steinbeck.* Ed. Kiyoshi Nakayama. Tokyo: Nan'un-do, 1986.

————. "The Vigilante." *The Long Valley.* 1938. New York: Penguin, 1986. 129-39.

————. *The Wayward Bus.* 1947. New York: Penguin, 1979.

————. *The Winter of Our Discontent.* 1961. New York: Penguin, 1986.

————. *Working Days: The Journals of The Grapes of Wrath.* Ed. Robert DeMott. New York: Viking Penguin, 1989.

————. *Zapata, The Little Tiger.* London: Heinemann, 1991.

Steinbeck, John and Edward F. Ricketts. *Sea of Cortez: A Leisurely Journal of Travel and Research.* New York: Viking, 1941.

Unpublished

Steinbeck, John. "Argument of Phalanx." Ms., c. 1935. The Bancroft Library, University of California, Berkeley.

————. "Case History." Ms., c. 1934. Steinbeck Research Center, San Jose State University, CA.

————. "The Days of Long Marsh." Ts., c. 1924. Houghton Library, Harvard University.

————. "East Third Street." Ts., c. 1925. Houghton Library, Harvard University.

———. "The Green Lady." Ts., c. 1928. Steinbeck Collection, Stanford University Library.

———. "The Leader of the People." Original manuscript, "*Tortilla Flat* notebook." Unpublished note. c. 1933-34. Harry Ransom Humanities Research Center, University of Texas, Austin.

———. Letter to George Albee. 1933. The Bancroft Library, University of California, Berkeley. [An edited version of this letter is published in *Steinbeck: A Life in Letters.* Ed. Elaine Steinbeck and Robert Wallsten. 1975. New York: Penguin, 1986. 83-84.]

———. Letter to George Albee. 5 Nov. [1937?] Harry Ransom Humanities Research Center, University of Texas, Austin.

———. Letter to James D. Brasch. 21 Sept. 1954. Steinbeck Collection, Stanford University Library.

———. Letters to Wanda Van Brunt. 12 Sept. 1948, 28 Sept. 1948, 16 Jan. 1949. Steinbeck Collection, Stanford University Library.

———. "Murder at Full Moon." Ms., c. 1930. Harry Ransom Humanities Research Center, University of Texas at Austin.

———. "The Nymph and Isobel." Ts., c. 1924. Houghton Library, Harvard University.

———. Notes in the notebook which contains the original manuscript of *In Dubious Battle.* C. 1935. Harry Ransom Humanities Research Center, University of Texas, Austin

———. Notes in the notebook which contains part of the original manuscript of *The Pastures of Heaven.* C. 1929-31. Steinbeck Collection, Stanford University Library.

———. "*Tortilla Flat* notebook." Ms., c. 1933-34. Harry Ransom Humanities Research Center, University of Texas, Austin.

Secondary Sources

Alexander, Stanley. "*Cannery Row:* Steinbeck's Pastoral Poem." *Steinbeck: A Collection of Critical Essays.* Ed. Robert Murray Davis. Englewood Cliffs, NJ: Prentice-Hall, 1972. 135-48.

Allee, W.C. *Animal Aggregations.* Chicago: U of Chicago P, 1931.

Astro, Richard. *John Steinbeck and Edward F. Ricketts: The Shaping of a Novelist.* New Berlin: U. of Minnesota P, 1973.

Astro, Richard and Tetsumaro Hayashi, eds. *Steinbeck: The Man and His Work.* Corvallis: Oregon State U P, 1971.

Beatty, Sandra. "Steinbeck's Play-Women: A Study of the Female Presence in *Of Mice and Men, Burning Bright, The Moon Is Down,* and *Viva Zapata!*" *Steinbeck's Women: Essays in Criticism.* Ed. Tetsumaro Hayashi. Steinbeck Monograph Ser. 9. Muncie, Ind.: The John Steinbeck Society of America, 1979. 7-16.

Beebe, William. *Galapagos, World's End.* New York: Putnam, 1924.

Benson, Jackson J. *Looking for Steinbeck's Ghost.* Norman: U of Oklahoma P, 1988.

————, ed. *The Short Novels of John Steinbeck*. Durham: Duke UP, 1990.

————. *The True Adventures of John Steinbeck, Writer*. New York: Viking, 1984.

Bergson, Henri. *Creative Evolution*. Trans. Arthur Mitchell. 1911. New York: H. Fulton, 1937.

Bloom, Harold, ed. *John Steinbeck*. New York: Chelsea House, 1987.

Boodin, John Elof. *Cosmic Evolution*. New York: Macmillan, 1925.

————. *A Realistic Universe*. 1916. 2nd ed. New York: Macmillan, 1931.

Bowlby, John. *Charles Darwin: A New Life*. New York: Norton, 1990.

Brent, Peter. *Charles Darwin: A Man of Enlarged Curiosity*. New York: Norton, 1981.

Briffault, Robert. *The Making of Humanity*. 1919. London: G. Allen & Unwin, 1928.

————. *The Mothers: The Matriarchal Theory of Social Origins*. New York: Macmillan, 1931.

Brown, Harold. "The Material World—Snark or Boojum?" *Journal of Philosophy* 22 (1925): 197-214.

————. "The Problem of Philosophy." *Journal of Philosophy* 17 (1920): 281-300.

Chaucer, Geoffrey. "The Prologue." *The Canterbury Tales*. *The Works of Geoffrey Chaucer*. Ed. F.N. Robinson. 1933. Cambridge, MA: Riverside, 1957.

Choi, Jin Young. "Steinbeck Studies in Korea." *John Steinbeck: Asian Perspectives*. Ed. Kiyoshi Nakayama, Scott Pugh, and Shigeharu Yano. Osaka: Osaka Kyoiku Tosho, 1992. 19-25.

Coers, Donald. *John Steinbeck as Propagandist:* The Moon Is Down *Goes To War*. Tuscaloosa: U of Alabama P, 1991.

Colp, R., Jr. *To Be an Invalid: The Illness of Charles Darwin*. Chicago: Chicago UP, 1977.

Davis, Robert Murray. "Steinbeck's 'The Murder.'" *Studies in Short Fiction* 14 (1977): 63-68.

DeMott, Robert. *Steinbeck's Reading: A Catalogue of Books Owned and Borrowed*. New York: Garland, 1984.

————. *"To a God Unknown."* *A Study Guide to Steinbeck*. Ed. Tetsumaro Hayashi. Metuchen, NJ: Scarecrow P, 1974. 187-213.

Ditsky, John. "Faulkner Land and Steinbeck Country." *Steinbeck: The Man and His Work*. Ed. Richard Astro and Tetsumaro Hayashi. Corvallis: Oregon State UP, 1971. 11-23

————. "Steinbeck, Bourne, and the Human Herd: A New/Old Gloss on *The Moon is Down*." *Rediscovering Steinbeck—Revisionist Views of His Art, Politics and Intellect*. Ed. Cliff Lewis and Carroll Britch. Lewiston, NY: Edwin Mellen, 1989. 177-91.

Enea, Sparky. *With Steinbeck in the Sea of Cortez*. (As told to Audrey Lynch.) Los Osos, CA: Sand River P, 1991.

Fensch, Thomas, ed. *Conversations with John Steinbeck*. Jackson, MS: UP of Mississippi, 1988.

————. *Steinbeck and Covici*. Middlebury, VT: Paul S. Eriksson, 1979.

Fiedler, Leslie. "Looking Back after 50 Years." *San Jose Studies* 16.1 (1990): 54-64.

Fontenrose, Joseph. *John Steinbeck: An Introduction and Interpretation*. New York: Holt, Rinehart, and Winston, 1963

Fowler, G. Herbert, and E.J. Allen. *Science of the Sea*. Oxford: Clarendon, 1928.

French, Warren. *John Steinbeck*. New York: Twayne, 1961.

Friedan, Betty. *The Feminine Mystique*. New York: Norton, 1963.

Gladstein, Mimi Reisel. "Steinbeck's Juana: A Woman of Worth." *Steinbeck's Women: Essays in Criticism*. Ed. Tetsumaro Hayashi. Steinbeck Monograph Ser. 9. Muncie, Ind.: The John Steinbeck Society of America, 1979. 49-52.

————. "Straining for Profundity: Steinbeck's *Burning Bright* and *Sweet Thursday*." *The Short Novels of John Steinbeck*. Ed. Jackson J. Benson. Durham: Duke UP, 1990. 234-48.

Goldsmith, Arnold. "Thematic Rhythm in *The Red Pony*." *Steinbeck: A Collection of Critical Essays*. Ed. Robert Murray Davis. Englewood Cliffs, N.J.: Prentice-Hall, 1972. 70-74.

Gonzales, Bobbi, and Mimi Gladstein. "*The Wayward Bus*: Steinbeck's Misogynistic Manifesto?" *Rediscovering Steinbeck—Revisionist Views of his Art, Politics and Intellect*. Ed. Cliff Lewis and Carroll Britch. Lewiston, NY: Edwin Mellen, 1989. 157-73.

Graubard, Mark. *Man the Slave and Master*. New York: Covici-Friede, 1938.

Hedgpeth, Joel W., ed. *The Outer Shores*. By Edward F. Ricketts. 2 vols. Eureka, CA: Mad River P, 1978.

————. "Philosophy on Cannery Row." *Steinbeck: The Man and His Work*. Ed. Richard Astro and Tetsumaro Hayashi. Corvallis: Oregon State UP, 1971. 89-129.

Hoffman, Arthur. "Chaucer's Prologue to Pilgrimage: The Two Voices." *Chaucer*. Ed. Edward Wagenknecht. New York: Oxford UP, 1959. 30-45.

Hughes, R. S. *John Steinbeck: A Study of the Short Fiction*. Boston: G. K. Hall (Twayne), 1989.

Huntington, Ellsworth. *Civilization and Climate*. 3rd ed. New Haven: Yale UP, 1924.

Kiernan, Thomas. *The Intricate Music*. Boston: Little Brown, 1979.

Levant, Howard. *The Novels of John Steinbeck: A Critical Study*. Columbia: U of Missouri P, 1974.

Lewis, R.W.B. "The Picaresque Saint." *Twentieth Century Interpretations: The Grapes of Wrath*. Ed. Robert Con Davis. Englewood Cliffs, NJ: Prentice-Hall, 1982. 144-49.

Lisca, Peter. "*The Grapes of Wrath*." *Steinbeck: A Collection of Critical Essays*. Ed. Robert Murray Davis. Englewood Cliffs, NJ: Prentice-Hall, 1972. 75-101.

————. *The Wide World of John Steinbeck*. New York: H. Wolff, 1958.

Lutwack, Leonard. "*The Grapes of Wrath* as Heroic Fiction." *The Grapes of*

Wrath: A Collection of Critical Essays. Ed. Robert Con Davis. Englewood Cliffs, NJ: Prentice-Hall, 1982. 63-75.

Malinowski, Bronislaw. *The Sexual Life of Savages.* 2 vols. New York: Horace Liveright, 1929.

Marks, Lester Jay. *Thematic Design in The Novels of John Steinbeck.* The Netherlands: Mouton, 1969.

Mead, Margaret. *Coming of Age in Samoa.* New York: Morrow, 1928.

———. *Growing Up in New Guinea.* New York: Morrow, 1930.

Mizener, Arthur. "Does a Moral Vision of the Thirties Deserve a Noble Prize?" *New York Times* 9 Dec. 1962: 4.

Moore, Harry Thornton. *The Novels of John Steinbeck: A First Critical Study.* Chicago: Normandie House, 1939.

———. *The Novels of John Steinbeck: A First Critical Study.* 1939. 2nd ed. Port Washington, NY: Kennikat, 1968.

Morsberger, Robert E. "Steinbeck's Happy Hookers." *Steinbeck's Women: Essays in Criticism.* Ed. Tetsumaro Hayashi. Steinbeck Monograph Ser. 9. Muncie, Ind.: The John Steinbeck Society of America, 1979. 36-48.

note. *Steinbeck Quarterly.* 4.2 (1971): 62-63.

Osborn, Henry Fairfield. Preface. *Galapagos, World's End.* By William Beebe. New York: Putnam, 1924. v-viii.

Owens, Lewis. *John Steinbeck's Re-vision of America.* Athens: U of Georgia P, 1985.

Perez, Betty L. *"In Dubious Battle." A Study Guide to Steinbeck: A Handbook to his Major Works.* Ed. Tetsumaro Hayashi. Metuchen, NJ: Scarecrow P, 1974. 47-68.

Review of *Cannery Row. Times Literary Supplement* 3 Nov. 1945: 521.

Review of *The Winter of Our Discontent. Newsweek* 26 June 1961: 96.

Review of *The Winter of Our Discontent. Times Literary Supplement* 7 July 1961: 413.

Prescott, Orville. Rev. of *Cannery Row. New York Times* 2 Jan. 1945: L-17.

———. Rev. of *The Wayward Bus. New York Times* 17 Feb. 1947: L-17.

———. Rev. of *The Pearl. New York Times* 24 Nov. 1947: L-21.

Ricketts, Edward F. *The Outer Shores.* 2 vols. *Part 1: Ed Ricketts and John Steinbeck Explore the Pacific Coast. Part 2: Breaking Through.* Ed. Joel W. Hedgpeth. Eureka, CA: Mad River P, 1978.

———. "The Tide." *The Outer Shores. Part 2: Breaking Through* 63- 68. Ed. Joel W. Hedgpeth, Eureka, CA: Mad River P, 1978.

Ritter, William Emerson. *The Unity of the Organism or The Organismal Conception of Life.* 2 vols. Boston: Gorham P, 1919.

Ritter, William Emerson, and Edna W. Bailey. "The Organismal Conception: Its Place in Science and Its Bearing on Philosophy." *University of California Publications in Zoology* xxxi (1931): 307-58.

Shively, Charles. "John Steinbeck: From the Tide Pool to the Loyal Community." *Steinbeck: The Man and His Work.* Ed. Richard Astro and Tetsumaro Hayashi. Corvallis: Oregon State UP, 1971. 25-34.

Smuts, Jan Christian. *Holism and Evolution.* New York: Macmillan, 1926.
"Steinbeck: Critical Thorns and a Nobel Laurel." *Newsweek* 5 Nov. 1962: 65.
Street, Webster. "Remembering John Steinbeck." *San Jose Studies* 19 (1975): 109-127.
Takamura, Hiromasa. "John Steinbeck's Dramatic World." *John Steinbeck: Asian Perspectives.* Ed. Kiyoshi Nakayama, Scott Pugh, and Shigeharu Yano. Osaka: Osaka Kyoiku Tosho, 1992. 91-101.
Timmerman, John H. *The Dramatic Landscape of Steinbeck's Short Stories.* Norman: U of Oklahoma P, 1990.
———. *John Steinbeck's Fiction: The Aesthetics of the Road Taken.* Norman: U of Oklahoma P, 1986.
Thompson, Ralph. Rev. of *Of Mice and Men. New York Times* 27 Feb. 1937: L-15.
———. Rev. of *The Red Pony. New York Times* 29 Sept. 1937: L-21.
Thurber, James. "What Price Conquest?" Rev. of *The Moon is Down. New Republic* 16 Mar. 1942: 370.
Watt, F.W. *John Steinbeck.* New York: Grove P, 1962.
Weeks, Donald. "Steinbeck Against Steinbeck." *John Steinbeck.* Ed. Harold Bloom. New York: Chelsea House, 1987. 7-17.
Wilson, Edmund. "John Steinbeck." *The Boys in the Back Room.* San Francisco: Colt P, 1941: 41-53.
Wolf, Naomi. *The Beauty Myth.* New York: William Morrow, 1991.
"Wrapped & Shellacked." *Time* 2 Nov. 1962: 41-42.
Wollenberg, Charles. Introduction. *The Harvest Gypsies.* Berkeley: Heyday Books, 1988.
Woodward, Robert H. "The Promise of Steinbeck's 'The Promise.'" *A Study Guide to Steinbeck's* The Long Valley. Ed. Tetsumaro Hayashi. Ann Arbor, MI: Pierian P, 1976. 97-103.

Unpublished

Albee, Richard. Letter to Robert DeMott. 10 Aug. 1979. John Steinbeck Research Center, San Jose State University, CA.
Astro, Richard. "Discovering Ricketts: Marine Science Enters the English Department." Steinbeck and the Environment Conference. Nantucket, 15 May 1992.
Brodwin, Stanley. "The Example of Darwin's *Voyage of the Beagle* in *The Log From the Sea of Cortez.*" Steinbeck and the Environment Conference. Nantucket, 15 May 1992.
French, Warren. "How Green was John Steinbeck?" Steinbeck and the Environment Conference. Nantucket, 16 May 1992.
Ricketts, Edward F. Letter to John Steinbeck. 25 Nov. 1947. The Bancroft Library, University of California, Berkeley.
Scardigli, Virginia. Letter to Brian Railsback. 23 Feb. 1993.
Steinbeck, Elaine. Personal interview. New York City. 2 Mar. 1993.

INDEX